AF839778

# Impact of UV and UV/$H_2O_2$ AOP on EDC Activity in Water

Awwa Research Foundation

## About the Awwa Research Foundation

The Awwa Research Foundation (AwwaRF) is a member-supported, international, nonprofit organization that sponsors research to enable water utilities, public health agencies, and other professionals to provide safe and affordable drinking water to consumers.

The Foundation's mission is to advance the science of water to improve the quality of life. To achieve this mission, the Foundation sponsors studies on all aspects of drinking water, including supply and resources, treatment, monitoring and analysis, distribution, management, and health effects. Funding for research is provided primarily by subscription payments from approximately 1,000 utilities, consulting firms, and manufacturers in North America and abroad. Additional funding comes from collaborative partnerships with other national and international organizations, allowing for resources to be leveraged, expertise to be shared, and broad-based knowledge to be developed and disseminated. Government funding serves as a third source of research dollars.

From its headquarters in Denver, Colorado, the Foundation's staff directs and supports the efforts of more than 800 volunteers who serve on the board of trustees and various committees. These volunteers represent many facets of the water industry, and contribute their expertise to select and monitor research studies that benefit the entire drinking water community.

The results of research are disseminated through a number of channels, including reports, the Web site, conferences, and periodicals.

For subscribers, the Foundation serves as a cooperative program in which water suppliers unite to pool their resources. By applying Foundation research findings, these water suppliers can save substantial costs and stay on the leading edge of drinking water science and technology. Since its inception, AwwaRF has supplied the water community with more than $300 million in applied research.

More information about the Foundation and how to become a subscriber is available on the Web at **www.awwarf.org.**

# Impact of UV and UV/$H_2O_2$ AOP on EDC Activity in Water

Prepared by:
**Karl G. Linden**
Department of Civil and Environmental Engineering, Pratt School of Engineering
Duke University, Durham, NC 27708-0287
and
**Seth W. Kullman**
Integrated Toxicology Program, Nicholas school of the Environment and Earth Sciences
Duke University, Durham, NC 27708-0287

Sponsored by:
**Awwa Research Foundation**
6666 West Quincy Avenue, Denver, CO 80235-3098

Published by:

Awwa Research Foundation

American Water Works Association

IWA Publishing

**DISCLAIMER**

This study was funded by the Awwa Research Foundation (AwwaRF). AwwaRF assumes no responsibility for the content of the research study reported in this publication or for the opinions or statements of fact expressed in the report. The mention of trade names for commercial products does not represent or imply the approval or endorsement of AwwaRF. This report is presented solely for informational purposes.

Copyright © 2007
by Awwa Research Foundation

All Rights Reserved

# CONTENTS

# LIST OF TABLES

# LIST OF FIGURES

# FOREWORD

The Awwa Research Foundation is a nonprofit corporation that is dedicated to the implementation of a research effort to help utilities respond to regulatory requirements and traditional high-priority concerns of the industry. The research agenda is developed through a process of consultation with subscribers and drinking water professionals. Under the umbrella of a Strategic Research Plan, the Research Advisory Council prioritizes the suggested projects based upon current and future needs, applicability, and past work; the recommendations are forwarded to the Board of Trustees for final selection. The foundation also sponsors research projects through the unsolicited proposal process; and various joint research efforts with organizations such as the U.S. Environmental Protection Agency, the U.S. Bureau of Reclamation, and the Association of California Water Agencies.

This publication is a result of one of these sponsored studies, and it is hoped that its findings will be applied in communities throughout the world. The following report serves not only as a means of communicating the results of the water industry's centralized research program but also as a tool to enlist the further support of the nonmember utilities and individuals.

Projects are managed closely from their inception to their final report by the foundation's staff and large cadre of volunteers who willingly contribute their time and expertise. The foundation serves a planning and management function and awards contracts to other institutions such as water utilities, universities, and engineering firms. The funding for this research effort comes primarily from the Subscription Program, through which water utilities subscribe to the research program and make an annual payment proportionate to the volume of water they deliver and consultants and manufacturers subscribe based on their annual billings. The program offers a cost-effective and fair method for funding research in the public interest.

A broad spectrum of water supply issues is addressed by the foundation's research agenda: resources, treatment and operations, distribution and storage, water quality and analysis, toxicology, economics, and management. The ultimate purpose of the coordinated effort is to assist water suppliers to provide the highest possible quality of water economically and reliably. The true benefits are realized when the results are implemented at the utility level. The foundation's trustees are pleased to offer this publication as a contribution toward that end.

David E. Rager
Chair, Board of Trustees
Awwa Research Foundation

Robert C. Renner, P.E.
Executive Director
Awwa Research Foundation

# ACKNOWLEDGMENTS


The authors of this report are indebted to the following water utilities and individuals for their cooperation and participation in this project:

- Southern Nevada Water Authority, Boulder Creek, NV, USA
- Greater Cincinnati Water Works, Cincinnati, OH, USA
- City of Wilmington, Wilmington, NC, USA
- City of Fayetteville, P.O. Hoffer WTP, Fayetteville, NC USA
- Harnett County Regional WTP, Harnett County, NC USA
- Sunset Ave Tar River WTP, Rocky Mount, NC USA
- Cary/Apex WTP, Cary/Apex NC USA
- Williams WTP, Durham NC, USA
- Joan Oppenheimer, Principal Scientist, MWH Applied Research Dept. Pasadena, CA, USA
- Jim Bolton, President, Bolton Photosciences Inc., Edmonton, AB Canada

Research carried out at Duke University could not have been possible without the help of students and laboratory staff. Mr. Erik Rosenfeldt, a doctoral candidate at the time of this research, was responsible for development of the UV testing protocols and performing the fundamental work on UV based destruction of the contaminants tested, as well as the analytical methods development. Dr. Pei-Jen Chen, a doctoral candidate at the time of this research, was responsible for establishing the capabilities for all the toxicological testing performed as part of this research as well as the impacts of water quality on the testing procedures and also performed many of the UV exposures. The PIs would also like to thank Dr. Sumpter and Dr. Soto for supplying YES-yeast and MCF-7 cell lines respectively and Dr. D. Schlenk for providing training and support for the YES assay. We are particularly thankful to Dr. D Hinton and his laboratory staff for their diligence in maintaining the Duke University Aquatics Facility and providing a constant source of fish and embryos for this study.

Direction provided by the Project Advisory Committee—including Dr. Bill Anderson, Mr. Mehul Patel, and Dr. Bill Sonzogni and AwwaRF Project Manager Alice Fulmer is greatly appreciated.

# EXECUTIVE SUMMARY

## INTRODUCTION

Recent advances in analytical chemistry as well as in-depth long-term ecological studies have indicated that chemicals in our water environment at very low levels may have significant impacts on human and environmental health. These chemicals are thought to originate mostly from industrial and pharmaceutical sources and are known to be present in many source waters subsequently used for public water supply. Unit processes utilized in water treatment are designed for removal of known pathogens and priority contaminants. However, with advances in analytical chemistry, environmental toxicology, and epidemiology, we are now able to detect and assess the toxicity of trace levels of contaminants in water. These advances have allowed scientists and regulators to recognize that many contaminants that are now of concern to public health and the environment are not removed using conventional water treatment processes. In addition, little information is available regarding the fate and distribution of these contaminants through advanced treatment processes currently under consideration for water treatment. One such group of contaminants is the endocrine disrupting compounds (EDCs).

The use of ultraviolet (UV) light for disinfection of water is a technique that is becoming increasingly common for water supplies. With the recent promulgation of the Long Term 2 Enhanced Surface Water Treatment Rule (LT2ESWTR), water utilities are under increasing regulatory pressure to decrease chlorinated disinfection byproducts and more effectively inactivate *Cryptosporidium spp*. These two objectives can be met through the implementation of UV light for disinfection as it is very effective for inactivating *Cryptosporidium* and can reduce the level of free chlorine use at water treatment plants. The application of UV into the treatment train invites the question – "What else can UV do for water treatment?" The use of UV for photolysis and oxidation of persistent chemical pollutants is one area that holds promise as a clean, effective technology for transforming environmental pollutants of concern. Because UV is likely to find its way into water treatment plants in the near future, the ability of UV technology to treat emerging chemical contaminants should be assessed. This research project was undertaken to evaluate the potential for UV light technology to degrade estrogenic environmental contaminants that may be found in raw or partially treated water supplies.

## PROJECT APPROACH

The hypothesis that guided this research was that UV and UV in combination with $H_2O_2$ treatment processes will be effective at degrading EDCs under certain operational conditions. The goal of this research was to assess, through use of bioassays and chemical analyses, the degradation, byproduct formation, and subsequent toxicity of EDCs following UV and UV/$H_2O_2$ oxidation treatment of water. To achieve this goal, the specific objectives of this research were to:

1) Perform controlled bench scale testing for UV and UV/$H_2O_2$ oxidation treatment processes to assess degradation of the known EDCs bisphenol-A, ethinyl estradiol, estradiol, and nonlyphenol, and subsequent formation of estrogenic byproducts over a range of UV operating points,
2) Assess changes in estrogenic activity utilizing both *in vitro* and *in vivo* based assays pre and post UV treatment at varying levels of UV application spanning from UV doses used in disinfection to those used for oxidation.
3) Develop information on degradation kinetics and hydroxyl radical rate constants for each compound tested.
4) Determine the treatment conditions under which the contaminant activity/toxicity is decreased to below detectible levels as evidenced through byproduct analysis and toxicity bioassays, indicating a completely treated water.

This work was carried out whereby contaminants were spiked into clean laboratory buffered water, one at a time, and exposed to UV alone and UV/$H_2O_2$ oxidation treatments at varying levels: from UV doses typically used in disinfection to doses used for contaminant degradation. Samples were analyzed for degradation of the parent compound. Treated samples with varying levels of byproducts were subjected to a suite of bioassays that served to indicate estrogenic activity and provide some insight on potential toxicity. The impact of different levels and types of UV treatment on estrogenic activity of the UV exposed contaminants was determined. Subsequently, contaminants were spiked into a series of raw and finished natural waters obtained from our utility partners. Degradation of the contaminants was assessed in these natural water matrices. Assessments of the estrogenic activity of these contaminants in the natural water matrices following UV treatment was also studied, as was the effects of water quality and mixed contaminants. Finally, the presence of estrogenic activity in natural waters at various water treatment plants in North Carolina was studied using an estrogenic screening assay.

## KEY FINDINGS

### Effectiveness of UV Treatment

The degradation of four EDCs: bisphenol-A, ethinyl estradiol, estradiol, and nonlyphenol was investigated via ultraviolet (UV) radiation photolysis and the UV/$H_2O_2$ advanced oxidation process (AOP). These EDCs have been detected at low levels in wastewaters and surface waters in both the United States and European countries, can cause adverse effects on humans and wildlife via interactions with the endocrine system, and thus must be treated before entering the public drinking water supply. For each EDC tested, direct UV photolysis treatment was assessed for both monochromatic low pressure (LP) UV lamps and polychromatic medium pressure (MP) UV lamps, and second order hydroxyl radical based rate constants were developed. These parameters were utilized to successfully model UV treatment of the EDCs in laboratory and natural waters. The polychromatic MP UV radiation source was more effective for direct photolysis EDC degradation as compared to conventional LP UV lamps emitting monochromatic UV 254nm radiation. However, in all cases the EDCs were more effectively degraded utilizing UV/$H_2O_2$ advanced oxidation as compared to direct UV photolysis treatment.

**Assessments Using Chemical Versus Bioanalytical Approaches**

The use of direct photolysis with LP and MP UV lamps and UV + hydrogen peroxide ($H_2O_2$) advanced oxidation process (AOP) was assessed for the degradation of bisphenol A (BPA) in laboratory water. Removal rates of BPA and formation of degradation products were determined by comparative chemical analysis and bioassays using the *in vitro* Yeast Estrogen Screen (YES) and *in vivo* Vitellogenin (VTG) with Japanese medaka fish (*Oryzias latipes*). Although UV alone did not effectively degrade BPA, UV in combination with $H_2O_2$ significantly removed BPA parent compound and aqueous estrogenic activity *in vitro* and *in vivo*. Removal rates of *in vivo* estrogenic activity were significantly lower than those observed *in vitro*, demonstrating differential sensitivities of these bioassays and that certain UV/AOP BPA metabolites may retain estrogenic activity. Furthermore, the UV/$H_2O_2$ AOP was effective for reducing larval lethality in treated BPA solutions, suggesting BPA degradation occurred and that the degradation process did not result in the production of acutely toxic intermediates. These differences highlight the differential sensitivities of *in vitro* and *in vivo* bioanalytical assays and emphasize the importance of employing mechanistically diverse assays for assessment of EDC activity in water treatment studies.

The transformation of the steroidal EDCs, 17-β-estradiol (E2) and 17-α-ethinyl estradiol (EE2) by direct UV photolysis and UV/$H_2O_2$ advanced oxidation was also studied from the perspective of the removal of estrogenic activity associated with the compounds. First, experiments were performed to link the oxidation of E2 and EE2 with subsequent reduction in estrogenic activity. No statistically significant difference between these rates was observed, implying that the oxidation products of E2 and EE2 are not as estrogenically active (as measured by the YES) as the parent compounds. With this knowledge, we utilized the YES to determine the removal of estrogenic activity and corresponding oxidation of E2 and EE2 at environmentally relevant concentrations (~3$\mu g\ L^{-1}$) in DI water and in natural waters. These results shown in Figure ES.1 implied that 1 log oxidation of these species occurred in less than 400 mJ $cm^{-2}$, implying these compounds are significantly degraded at much lower UV doses than previously thought, when considered at environmentally relevant levels. Finally, the steady state OH radical model was used successfully to predict oxidation of EE2 in DI water, but significantly under-predicted oxidation rates in natural waters.

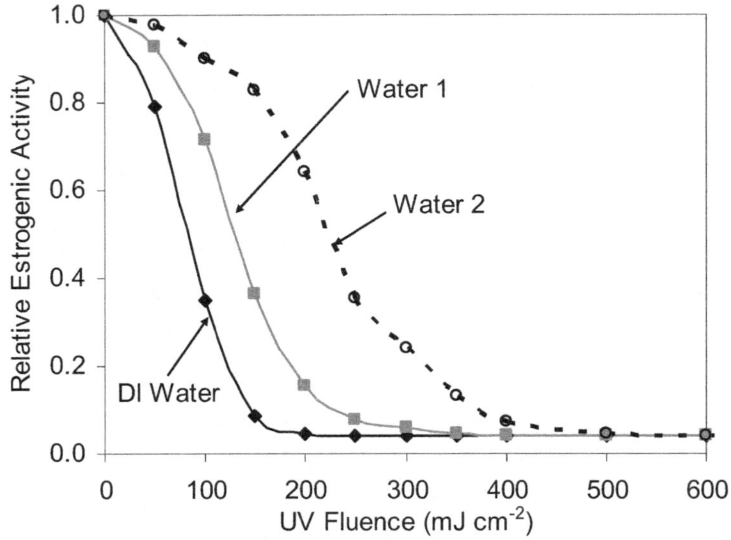


**Figure ES.1 Reduction of Estrogenic activity using MP UV + 5 mg $L^{-1}$ $H_2O_2$ AOP to oxidize 3μg $L^{-1}$ EE2 in DI water and 2 natural waters.**

Natural variations in water matrices and mixtures of EDCs in the environment may confound analysis of treatment efficiencies. Bioanalytical assessments of residual estrogenic activity in treated water were used to evaluate the performance of UV based advanced oxidation (UV/AOP) for estrogenic contaminants in natural water and in mixtures. Four EDCs including $E_2$, $EE_2$, BPA and nonylphenol (NP) were spiked individually or as a mixture at μg $L^{-1}$-ng $L^{-1}$ in laboratory or natural river water. Removal rates of estrogenic activity were quantitatively evaluated by *in vitro* Yeast estrogen screen (YES) and *in vivo* Vitellogenin (VTG) assays. UV/AOP treatment was capable of decreasing *in vitro* and *in vivo* estrogenic activity, however, at UV fluences up to 2000 mJ $cm^{-2}$ in the presence of 10 mg $L^{-1}$ $H_2O_2$ some *in vivo* estrogenic activity remained. These results suggest that degradation of EDC mixtures at environmentally relevant concentration is more complex than that observed with individual compounds. Removal rates of *in vitro* estrogenic activity of the EDC mixtures were lower than those observed for single compounds, and slower in natural waters, likely due to lower steady-state concentrations of hydroxyl radicals (•OH) in the presence of the •OH scavengers from the water matrix and EDC mixture.

**Estrogenic Activity at Water Treatment Plants**

As part of the natural water evaluations, it was noted that some waters that were collected at the inlet to WTPs contained some background estrogenic activity. These findings prompted a side project to investigate the presence of estrogenic EDCs as detected using the YES assay at a number of water treatment plants around the State of North Carolina in different watersheds as illustrated in figure ES.2. There turned out to be no observed estrogenic activity as measured with the YES in the unconcentrated water samples collected from the utilities studied in this project. When samples were concentrated, they yielded some interesting observations, including an apparent correlation between the location of the water source within an urban to suburban area and increased estrogenic activity. Utilities with sources located in more rural or small town settings did not show positive responses to the YES.

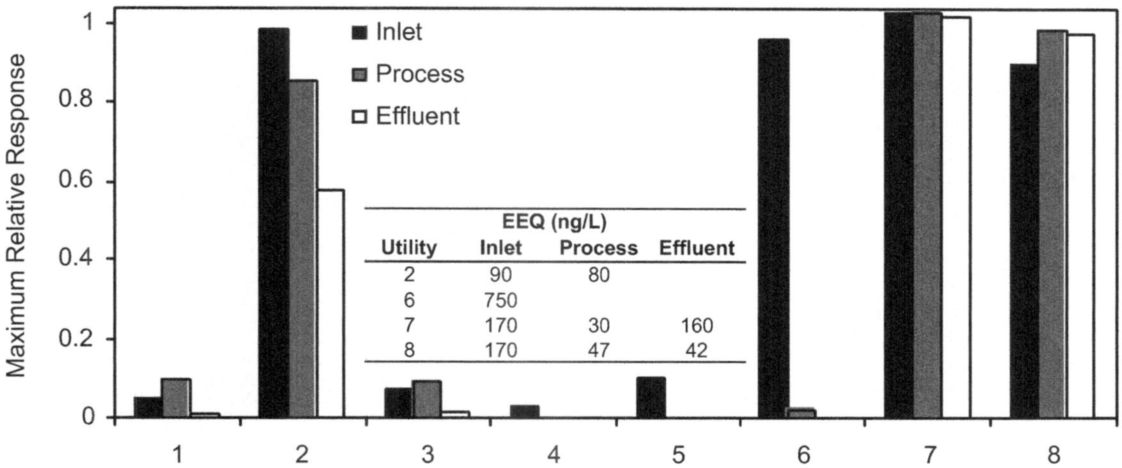


| Utility | Inlet | Process | Effluent |
|---|---|---|---|
| 2 | 90 | 80 | |
| 6 | 750 | | |
| 7 | 170 | 30 | 160 |
| 8 | 170 | 47 | 42 |

**Figure ES.2 Maximum relative YES response corresponding to 50x concentration of the water samples**

Estrogenic activity was found associated with suspended solids from the inlet waters of all the utilities tested with the exception of one facility. This was expected due to the hydrophobic nature of many xenoestrogenic compounds, and a mass-based estrogen equivalent concentration for the solids (mg $kg^{-1}$) was calculated.

# CHAPTER 1
# INTRODUCTION

## BACKGROUND

Recent advances in analytical chemistry as well as in-depth long-term ecological studies have indicated that chemicals in our water environment at very low levels may have significant impacts on human and environmental health. These chemicals are thought to originate mostly from industrial and pharmaceutical sources and are known to be present in many source waters subsequently used for public water supply. Unit processes utilized in water treatment are designed for removal of known pathogens and priority contaminants. However, with advances in analytical chemistry, environmental toxicology, and epidemiology, we are now able to detect and assess the toxicity of trace levels of contaminants in water. These advances have allowed scientists and regulators to recognize that many contaminants that are now of concern to public health are not removed using conventional water treatment processes. In addition, not much is known about the fate of these contaminants through emerging advanced treatment processes currently under consideration for water treatment. One such group of contaminants is the endocrine disrupting compounds (EDCs).

The use of ultraviolet (UV) light for disinfection of water is a technique that is becoming increasingly common for water supplies. With the recent promulgation of the Long Term 2 Enhanced Surface Water Treatment Rule (LT2ESWTR), water utilities are under increasing regulatory pressure to decrease chlorinated disinfection byproducts and more effectively inactivate *Cryptosporidium spp*. These two objectives can be met through the implementation of UV light for disinfection, as it is very effective for inactivating *Cryptosporidium* and can reduce the level of free chlorine use at water treatment plants. The application of UV into the treatment train invites the question – "What else can UV do for water treatment?" The use of UV for photolysis and oxidation of persistent chemical pollutants is one area that holds promise as a clean, effective technology for transforming environmental pollutants of concern. Because UV is likely to find its way into water treatment plants in the near future, the ability of UV technology to treat emerging chemical contaminants should be assessed. This research project was undertaken to evaluate the potential for UV light technology to destroy estrogenic environmental contaminants that may be found in raw or partially treated water supplies.

### Significance to Water Utilities

The extent to which EDCs are contaminating our drinking water supplies is currently being established and future applications of methods to assess toxicity of drinking water supplies and treatment processes that can remove these contaminants will be paramount. Interest has developed in assessing the potential of conventional and advanced water treatment processes to treat/remove select EDCs. Technological advances in water treatment processes including UV and UV based oxidation processes are now showing promise for treating emerging water contaminants such as N-nitrosodimethylamine (NDMA) and pharmaceuticals. It is predicted that these treatments will additionally be effective for the removal of numerous EDCs. When coupled together with highly sensitive bioassays, these treatment based research efforts will provide complementary information on achievable levels of contaminant removal and potential byproducts from potable water sources, and their subsequent toxicity.

This research is relevant to the water industry in numerous ways. The findings provide important information for regulators, utilities, and consulting engineers on the effectiveness of UV based processes and increase the knowledgebase for this water treatment technology. A perspective on the relationship between the level of UV treatment needed for disinfection as compared to that necessary for EDC contaminant degradation helps designers engineer UV systems to achieve multiple endpoint treatments. The scientific community will benefit from the generation of fundamental rate constants information for UV photolysis and hydroxyl radical reactions with these contaminants.

Perhaps the two most important outcomes of this work will be new knowledge and understanding relating to the byproducts that result when UV and UV oxidation are used to treat water containing EDCs and the subsequent toxicity of the water and byproducts that may be generated. This outcome of the work is one aspect that makes the research unique among recent treatment based projects. We report the level of UV treatment necessary that results in a complete decrease in toxicity and compare various UV treatments on the efficacy for decreasing the endocrine based toxicity of the waters tested. This is an important step beyond the traditional mode of just testing for disappearance of the parent compound. As part of the toxicity evaluations, a comprehensive testing protocol was developed that could serve as a benchmark for utilities and regulators in assessing toxicity. The standardized testing battery utilized in this research could be used in future assessments of water treatment technologies in evaluating their removal of EDCs.

The findings from this research should be readily usable by the utility and research community. One important study to follow this project is addressing the issue of contaminant mixtures. We were able to begin some of this work as presented below but follow-up in this area will be important. Another area for more research would be in evaluating some emerging UV technologies and novel UV sources. Finally, pilot or full scale testing would be important to establish operating parameters to ease treatment plant implementation of the research.

## UV Technologies for Disinfection

There are four major types of UV technologies that are used for drinking water treatment: low-pressure (LP) lamps, low pressure high-output (LPHO) lamps, medium-pressure (MP) lamps, and pulsed (P) lamps. These lamp types differ in the method and intensity of emitted UV light at each wavelength. While LP lamps are effective at disinfecting target microorganisms, there are several disadvantages that make them impractical for some treatment facilities. These include high lamp numbers and large system footprints for facilities servicing more than 10 million gallons per day (mgd) (Cairns 1996, Kwan et al. 1996). High maintenance and labor costs may also be associated with cleaning large numbers of UV lamps with sleeve fouling. Consequently, there may be interest in evaluating the use of higher irradiance UV sources such as LPHO-, MP- and P-UV lamps for larger scale drinking water applications. To date, P-UV has been in the research stage only, thus was not evaluated in this project. A description of low and medium pressure UV sources is provided below.

### *Low Pressure Systems*

UV light is emitted as a result of current flow through the mercury vapor between the electrodes of the lamp. The most commonly used UV lamps (low pressure mercury vapor) produce the majority of their UV output at 253.7 nm, a wavelength which is very close to the 260 – 265 nm wavelengths which are most effective in the inactivation of microbes. LPHO-UV lamps also emit monochromatic radiation principally at 253.7 nm, but at a slightly higher intensity. The germicidal UV fluence (*H'*) for these lamps is typically calculated as the product of the volume averaged incident irradiance (*E*, mW $cm^{-2}$) at 253.7 nm and the time of exposure (*t*, seconds) resulting in units of millijoules per square centimeter (mJ $cm^{-2}$) for UV fluence.

***Medium-Pressure Systems***

MP-UV lamps are mercury-arc lamps that operate at up to a million times the pressure and at much higher temperatures than LP-UV lamps. The increased pressure causes emission of significantly higher intensity radiation and a broader emission lines. As a result, the emission from an MP lamp becomes quasi-continuous and polychromatic over the range from the far UV (185 nm) to the infrared (1,367 nm). A small fraction of the polychromatic radiation (approximately 15 to 25 percent) emitted is in the germicidal range (200-300 nm). Due to this smaller fraction, MP lamps are much less germicidally effective than LP lamps. However, MP-UV lamps can emit approximately 60 to 600 times the germicidal irradiance of the LP-UV lamps. Two of the strongest emissions in this range are at 266 and 256 nm. This higher emittance translates into a comparatively smaller footprint for the MP-UV disinfection system (significantly fewer lamps for equivalent effect) and potentially lower operations and maintenance costs.

## LITERATURE REVIEW

### Endocrine Disruption in Humans and Wildlife

The group of chemicals known as endocrine disrupting chemicals (EDCs) can cause adverse effects on human and wildlife via interactions with endocrine systems. These contaminants may act to mimic hormones, to antagonize normal hormones, to alter the pattern of synthesis and metabolism of natural hormones or to modify hormone receptor levels (Sonnenschein and Soto, 1998). For example, the *in vivo* deleterious effects of exposure to such compounds may be seen as altered gonadal morphology and function, alteration in external genitalia, alteration in sex ratios, abnormal development, or as manifestations of neural and reproductive toxicity (Colborn, et al., 1993).

While findings of endocrine disruptive effects preceded definitive information on chemicals and their etiologic roles, exposure to numerous pesticides, industrial chemicals and natural products demonstrate causal relationship. EDCs of anthropogenic origin (xeno-estrogens) that are of concern in water and wastewater include pesticide residues (e.g., DDT, endosulfan, methoxychlor), polychlorinated biphenyls, dioxin, alkylphenols (e.g., nonlyphenol), plastic additives (e.g. bisphenol A, diethyl phthalate), polycyclic aromatic hydrocarbons (e.g., benzo(a)pyrene, benz(a)anthracene, chrysene) and pharmaceutical hormones (e.g., ethynyl estradiol) (USEPA, 2001). Natural compounds in the environment such as phyto-estrogens also may be of concern. Known mechanisms of action include: estrogen mimics (Welch, et al., 1969) as mentioned above, antiandrogens (Kelce et al., 1995) and /or thyroid hormone mimics (Morse et al., 1996).

Worldwide, the number of compounds studied and followed by regulators comprises only 0.1% of the more than 150,000 chemicals introduced by humankind. The EDCs require analysis and determination of their fate and transport, and potential for release. Recent research has demonstrated that many of these contaminants are found in the environment at concentrations sufficient to cause deleterious effects in wildlife. Wildlife species have provided strong additional *in vivo* evidence of endocrine disruption. Reports of adverse effects have emanated from investigations of reptiles, birds, and fish, all egg-laying animals. For example, one of the earliest involved seagulls along the California coast displaying abnormal sex differentiation associated with exposure to o,p' DDT (Fry and Toone, 1981). Later, bald eagles of the Great Lakes exhibited poor embryonic survival due to eggshell thinning and cracking caused by exposure to chlorinated organics (Weimeyer,et al., 1984). In the late 1980s, in an alligator population in Lake Opopka, FL, numbers of males sharply declined after chronic exposure to DDT, which caused feminization and the loss of fertility (Guillette, et al., 1994).

Alkyl phenols, formed during the biological degradation of detergents in sewage, and detected in drinking water (Clark et al., 1992) were implicated in the feminization of male fish in several rivers in the United Kingdom (Purdom et al., 1994). Jobling et al. (1996) further reported retardation in spermatogenesis in rainbow trout exposed to the same estrogenic compound, indicating its potential to diminish reproduction. Sumpter and Jobling (1995) reported that vitellogenin, the yolk precursor normally restricted to females, was induced in male or immature rainbow trout deployed near the effluent of 15 different sewage treatment plants in England and Wales. This illustrates that egg-laying animals are particularly good indicators of estrogenic contaminants in water. Males, possessing estrogen receptors, apparently respond by producing vitellogenin synthesis after exposure to estrogen mimics. To this extent, Folmar et al (1996) showed induction of hepatic vitellogenin and reduced serum testosterone concentrations in wild male carp (*Cyprinus carpio*) sampled near a large sewage treatment plant. In a national survey of surface waters, the U.S. Geological Survey (Goodbred et al, 1997) reported induction of vitellogenin in common male carp from a variety of U.S. streams. And in wild male fishes, induction of vitellogenin and of an intersex condition in gonads (gonads with both male and female germ cells) were identified following exposure to sewage effluents (Toomey et al., 1999; Jobling et al., 1998; Folmar et al 1996). Feminization of fish in domestic sewage was also attributed to ethynyl estradiol, a synthetic estrogen used in birth control pills (Sonnenschein and Soto, 1998). Levels of ethynyl estradiol in wastewaters have been demonstrated in the range 1-3 ng $L^{-1}$ (Belfroid et al., 1999) and are likely to increase in basins that are subject to agricultural runoff from animal husbandry operations where hormone treatment for increased productivity is on the rise. Bisphenol A has been found in waters stored in polycarbonate bottles and in canned foods (Brotons et al., 1995). Pesticide residues are found in many natural waters due to agricultural runoff and irrigation return waters as well in sewage from industrial and municipal effluents. The wetting agents in many pesticide applications are EDCs.

Most recently a US Geological Survey (USGS) study was published that detailed the occurrence of some of the most prevalent organic wastewater contaminants, including pharmaceuticals and hormones (Kolpin et al., 2002) in US streams. Among the hundreds of contaminants detected were bisphenol-A (in 41% of all samples), ethynyl estradiol (in 16% of all samples) and nonylphenol (in 50% of all samples). Recently in the news the popular press has picked up on the estrogenic effects of contaminants in water, concerning readers about feminization of fish, decreasing male sperm counts, and the effects of atrazine on mutations in frogs (Lean and Sadler, 2002; CNN, 2002; Hayes et al., 2002). The American Water Works

Association recently hosted a conference on Endocrine Disrupters and the Water Industry (April 2002). Presentations at this well attended symposium ranged from detection methods, attributes of different toxicity testing regimes, and treatment. Some of the work presented there is referenced elsewhere in this proposal.

**Water Quality Toxicity Testing**

The Federal Water Pollution Control Act, commonly known as the Clean Water Act (CWA), was enacted in 1972 with the objective of "restoring the chemical, physical, and biological integrity of the Nation's waters." Several goals and policies were established in the act including eliminating the discharge of pollutants and maintaining water quality sufficient for the protection and propagation of fish, shellfish, and wildlife. In 1987, the CWA stated that "no toxics in toxic amounts" shall be discharged into U.S. waters (USEPA 1991). In the 29 years since the CWA was enacted, the U.S. Environmental Protection Agency (EPA), through the National Pollutant Discharge Elimination System (NPDES) permitting program, has made significant progress toward controlling toxic discharges and implementing water quality standards. Industries and municipalities must comply with these federal standards by conducting whole effluent toxicity testing (WET) using the United States Environmental Protection Agency (USEPA) three species assays (USEPA, 1993a; 1993b; 2000). Although WET testing has led to a vast improvement of surface water quality, the effort is directed at acute toxicity. Endemic organisms however, are now exhibiting chronic toxicities including developmental impairment, altered sexual differentiation, and impaired reproduction from exposure to compounds undetected by these analyses including numerous endocrine disruption chemicals such as selenium, residues of the human birth control agent, ethynyl estradiol, and agricultural pesticides and industrial wetting agents.

To address this issue the USEPA has issued mandates initiating testing programs to determine which chemicals act to disrupt the endocrine system. The Food Quality Protection Act (1996) and the Safe Drinking Water Act (1996) and an Endocrine Disruptor Screening and Testing Advisory Committee (EDSTAC) has recommended use of cell assays and QSAR to *screen* industrial and agricultural chemicals for endocrine-disrupting effects (Ankely, G. et al. 1998). Most of the tests currently being used to screen EDC's and reproductive toxicity involve rapid *in vitro* and *in vivo* assays. Based on the premise that EDC's operate via similar modes of action between species (i.e. estrogen receptor and androgen receptor agonists and antagonists), rapid *in vitro* and *in vivo* assays have been developed to ascertain if a given compound could potentially affect estrogen/androgen/thyroid regulated biological processes. These assays include hormone receptor binding assays, reporter systems to measure transcriptional activation, and the proliferation of MCF-7 breast cancer cells (E-screen) (Garcia-Reyero et al, 2001; Jorgensen et al, 2000; Soto. et al, 1995). Additionally since reproductive proteins including vitellogenin (Vtg) and choriogenin (Chg) induction are under the control of estrogen in oviparous organisms, it has been broadly used as a biomarker of exposure for xenoestrogens (Arukwe et al, 2001; Arukwe et al, 1997). Fish hepatocytes have been used to study estrogen-like effects of contaminants. Vitellogenin (Vtg) in the culture media is taken as an indicator of estrogenicity (Anderson, et al., 1999).

There are significant limitations of the *in vitro* and *in vivo* methods noted above and use of any one method may not accurately depict a complete exposure and/or effect scenario. First, *in vitro* methods cannot account for the effects of EDCs on physiological processes. The role of xenobiotic metabolizing enzymes in biotransformation might not be considered; the parent toxicant could be made more or less estrogenic as a consequence of metabolism. The half-life ($t_{1/2}$) of the xenobiotic, which could have significant effects on its toxicity, cannot be accounted for. Many lipophilic compounds are stored in adipose tissue. As adipose tissue turns over, the animal is potentially exposed to small but potentially significant amounts of the xenobiotic(s). Single or multiple exposures in a culture dish cannot account for all of these important physiological effects. Additionally Vtg and Chg, although excellent markers for estrogenic effects, may not be useful for assessing non-estrogenic effects.

Over the past 10 years the laboratory group of the Co-PI's (Drs. Hinton and Kullman) has developed highly standardized *in vivo* and *in vitro* models of chronic toxicity and endocrine disruption using the medaka fish (Teh and Hinton, 1998; Cooke and Hinton, 1999; Koger et al., 2000). These models operate under standardized procedure in which environmental conditions are rigorously controlled producing a calibrated exposure scenario. In-vivo assays are specific for determination of reproductive and developmental abnormalities associate with EDC exposure including fertility, fecundity, maturation, hatching success, sex ratios, intersex (development of ovotestis) and embryo development. Exposure can be acute or chronic, and effect of reproductive impairment carried throughout subsequent F1/F2 generations. The medaka *in vivo* model has additionally been used in a number of studies focusing on endocrine disruption and reproductive toxicity, and has been proposed as an aquatic model for this type of work by numerous labs (Papoulias et al., 1996; Arcand-Hoy and Benson, 1998; Koger et al., 2000; Papoulias et al., 2000). In addition the Co-PI's have developed standardized procedures incorporating microarray analysis for the identification of novel genes and gene expression profiles associated with endocrine modulation in aquatic vertebrates. This new technology has not previously been applied to the detection of EDCs in the aquatic environment and will significantly accelerate the identification of novel markers of endocrine toxicity and determination of ED exposure and effect.

**Human Consequences of EDC's in the Environment**

Over the last ten years concern has been developed regarding the possible harmful effects of environmental chemicals capable of modulating the endocrine system. This concern is directed at numerous observed and suspected adverse effects on reproduction in both humans and wildlife. Numerous field and laboratory investigations have established a well-documented link between exposure to endocrine modifying compounds and reproductive abnormalities in wildlife species. Linkage of exposure to endocrine disruptors and potential adverse reproductive effects in humans has been more difficult to establish. Most noteworthy however has been the linkage between sons and daughters born to women treated with diethylstilbestrol (DES). Reports have additionally documented a decrease in sperm quality and counts, an increase in incidence of male genital abnormalities, altered sex ratios, premature development, and an increase in incidence of specific reproductive cancers including testicular, prostate, breast, and ovarian (Solomon and Schettler, 2000). Fetal development may be particularly sensitive to the effects of endocrine disruption. Small fluctuations in highly regulated hormonal cues may result in permanent physiological changes that are not observed in adults exposed at similar levels. Epidemiological studies have found associations between exposure to specific pesticides or industrial chemicals and altered hormone levels, impaired sexual function and other reproductive abnormalities (Solomon and Schettler, 2000). It must be noted however that no causal link between environmental exposure to endocrine modulating chemicals and these reproductive impairments, although suspected, has not yet been established.

### Treatment Processes for Removal of EDCs

Although conventional biological treatment processes have been shown to be effective at reducing levels of EDCs in sewage (Tanaka et al., 2001) the low levels of these contaminants in wastewater effluents are still a major concern for the environment and downstream endusers because EDCs exert physiological effects at very low concentrations. Garcia-Reyero et al. (2001) reported that the efficiency of biological based sewage treatment plants for degrading estrogenic compounds varied significantly and that EDCs were prevalent in river samples taken. These river sources are often the intakes for a water treatment process. Thus it is necessary to evaluate conventional and advanced water treatment processes and identify those that are effective at degrading these contaminants, and rendering them non-toxic.

One process that has gained increasing interest in the water industry over the past few years is ultraviolet (UV) light treatment. Since UV was found to be effective against the persistent pathogen *Cryptosporidium parvum* (Clancy et al, 2000; Shin at al, 2001) there has been a surge in research related to UV processes in water treatment. It is now known that UV is effective for disinfection of a wide range of health related waterborne pathogens. Thus the US EPA is accepting the use of UV light to meet some of the water quality goals associated with the long term 2 enhanced surface water treatment rule (LT2ESWTR). The use of UV has also extended to contaminant degradation in water. Recent studies on degradation of MtBE, 1,4-dioxane, NDMA, and atrazine (Stefan and Bolton, 1998; Stefan et al., 2000; Cater et al., 2000; Sharpless and Linden, 2003; Stefan et al., 2001) as well as taste and odor contaminants are an indication of the rise in interest in UV processes for contaminant degradation. In fact for persistent contaminants such as NDMA, the UV photolysis process is one of the only known effective treatments and was recently selected for treatment of NDMA and 1,4-dioxane in Orange County California for the Groundwater Replenishment System (OCWD, 2002).

A titanium dioxide ($TiO_2$) photocatalysis advanced oxidation process has recently been shown to completely degrade the EDC bisphenol A (BPA) to mineralization products (Ohko et al., 2001). The successful degradation by the oxidizing $TiO_2$ process indicates that other advanced oxidization processes that produce hydroxyl radicals (such as $UV+H_2O_2$) may also be promising technologies and would likely be more efficient at such degradation. Zwiener and Frimmel (2000) also found advanced oxidation was effective in removal of pharmaceutical wastes where as ozone alone was not as effective.

Further background material and literature is presented in individual chapters, relevant to the topics addressed therein.

## RESEARCH SCOPE AND OBJECTIVES

The hypothesis that guided this research was that UV and $UV/H_2O_2$ treatment processes will be effective at degrading EDCs under certain operational conditions. The goal of the research was to assess, through use of bioassays and chemical analyses, the degradation, byproduct formation, and subsequent toxicity of EDCs following UV and UV oxidation treatment processes of water. To achieve this goal, the specific objectives of this research were:

1) Perform controlled bench scale testing for UV and UV oxidation treatment processes to assess degradation of the known EDCs bisphenol-A, ethinyl estradiol, estradiol, and nonlyphenol, and subsequent formation of estrogenic byproducts over a range of UV operating points,
2) Assess changes in toxicity utilizing both *in vitro* and *in vivo* based assays pre and post UV treatment at varying levels of UV application spanning from UV doses used in disinfection to those used for oxidation.
3) Develop information on degradation kinetics and hydroxyl radical rate constants for each compound tested.
4) Determine the treatment conditions under which the contaminant toxicity is decreased to below detectible levels as evidenced through byproduct analysis and toxicity bioassays, indicating a completely treated water.
5) Evaluate mixtures of contaminants at environmentally relevant concentrations and the effects of water quality on UV based treatments.

## KEY PROJECT QUESTIONS

This research focused on a number of Key Questions that formed the basis of our experimental plan. These included:

- How do the UV doses necessary for EDC degradation compare to those needed for UV disinfection?
- How does water quality affect UV contaminant degradation treatment processes?
- How does UV light from different sources affect treatment of different EDCs?
- To what extent does UV-based advanced oxidation improve treatment for various contaminants compared to direct UV photolysis?
- What types of bioassays can be practically used for assessment of estrogenic activity?
- How do mixtures of contaminants impact UV treatment and subsequent estrogenic activity?
- Does the required UV treatment level differ for contaminants at environmental concentrations as compared to higher spiked levels?

## PROJECT TEAM

This project was a joint effort of a number of two different research groups under the coordination of Professor Karl G. Linden at Duke University. The environmental engineering group at Duke University was directly responsible for carrying out the bench scale UV work for all the EDC contaminants, the analytical chemistry assessments, and evaluating each UV technology. Dr. Seth Kullman headed up the research group from the Integrated Toxicology Program at Duke University that led the work on the bioassays for evaluating estrogenic activity and treatment effectiveness. Mr. Erik Rosenfeldt and Dr. Pei Jen Chen both worked on the project as doctoral students in Environmental Engineering at Duke.

# CHAPTER 2
# MATERIALS AND METHODS

This chapter describes the major methods and techniques used throughout the research project. Where necessary, any changes from these procedures or specific protocols and experimental designs are noted in the subsequent individual chapters.

## REAGENTS, CONTAMINANTS, AND WATERS

The EDCs used in this study included bisphenol-A (BPA, Aldrich Chemical Company, Milwaukee, WI), ethinyl estradiol (EE2, Steraloids, Inc., Newport, RI) and estradiol (E2, Steraloids, Inc., Newport, RI), and 4-Nonylphenol (NP, Supelco, Bellefonte, PA) all purchased as solids. The chemicals used for the mobile phase of HPLC detection included HPLC grade acetonitrile (Burdick and Jackson, Muskegon, MI), and water purchased from Malinkrodt Chemical Company (St. Louis, MO). Lab water was deionized and purified water taken from a laboratory water purifying system (Hydro Services and Supplies, Inc., Research Triangle Park, NC). A synthetically produced model water, which simulates the composition, pH, and properties of natural water, was used for the fundamental UV degradation studies. All concentrations of each component within the water were thus known and controllable allowing for a reproducible water matrix to be used over time. The main constituents within the water are displayed in Table 2.1. In subsequent natural water studies, waters were collected from various locations described in later sections.

## CHEMICAL ANALYSIS EQUIPMENT AND METHODS

A Varian Pro Star High Performance Liquid Chromatography (HPLC) workstation (Walnut Creek, CA), with a model 330 Polychromatic Diode Array detector was used to detect the concentration of EDCs in water. A C-18 reverse phase column served as the stationary phase, with 1 mL $min^{-1}$ of eluent consisting of an acetonitrile (ACN) to water ratio of 55%:45% for BPA, 60%:40% for EE2, and 50%:50% for E2. These HPLC methods were adapted from the literature for EE2, BPA, and E2 (Huber et al. 2003, Ohko et al. 2001, Ohko et al. 2002). For NP detection, a C-8 column was utilized with 1 mil $min^{-1}$ of 50%:50% ACN:$H_2O$ eluent, adapted with changes from Cheng and Ding, 2002. Water quality parameters were measured using standard methods. Total organic carbon (TOC) was measured in accordance with Standard Method 5310 A (combustion and detection method) (Tekmar Dohrmann Apollo 9000 Total Carbon analyzer). Alkalinity was measured according to Standard Methods 2320 B. Residual $H_2O_2$ was measured using the $I_3^-$ method (Klassen et al. 1994). pH was measured using an electronic pH meter (Cole Parmer pH 100 series), calibrated daily using pH 4, 7, and 10 buffers. UV absorbance (200-300 nm) of the spiked test water was measured in a spectrophotometer (Cary 100 bio-spectrophotometer, Varian, Houston, TX).

### UV Apparatus

Irradiation by a 1 kW medium pressure (MP) UV lamp (Hanovia Co., Union NJ) was performed using a bench-scale collimated beam UV apparatus provided by Calgon Carbon Corporation (CCC, Pittsburgh, PA) modified by Duke University. Irradiation by a low pressure

(LP) UV lamp was performed using a bench-scale collimated beam UV apparatus consisting of four 15 W LP lamps (ozone-free, General Electric #G15T8). Figure 2.1 shows a typical quasi-collimated beam UV system consisting of the light source, a "collimating" device, a support stand, and an exposure control system comprised of a manual (LP UV) or automatic (MP UV) shutter. A UV radiometer and detector (International Light Inc., Model 1700/SED 240/W) calibrated at 2 nm intervals in the range of 200 to 400 nm was used to measure UV irradiance at the surface of the test water. Relative output spectra of the LP and MP UV sources used in this study are presented in Figure 2.2.

**Fluence Calculation**

UV fluence (mJ $cm^{-2}$) was calculated as the average irradiance multiplied by the exposure time. The average UV irradiance in a completely mixed sample was determined from the incident irradiance, UV absorbance, and sample depth using an integrated form of the Beer-Lambert law (Bolton and Linden, 2003). Incident UV irradiance (mW $cm^{-2}$) was measured with a radiometer and detector at the surface of the liquid suspension. Samples were exposed to UV light for specific time intervals using a shutter system above the reaction vessel. The exposure time (seconds) necessary was determined by dividing the desired UV fluence by the average UV irradiance. For the LP UV source, the fluence was calculated as the radiation emitted at 253.7 nm. For the broadband MP UV source, the fluence was calculated as the total UV output in the 200 to 300 nm region.

**Irradiation and Sampling Procedure**

Unless otherwise noted, water samples were spiked with EDCs, and vacuum filtered through a Gelman Laboratory FP-450 Vericel Membrane Filter (0.45 μm pore size) to remove solid particles for protection of the HPLC column. A sample of the filtrate was taken and analyzed by HPLC to determine the concentration after filtration. Typically, 100 mL samples were measured into 70x50 flat bottom Pyrex petri dishes, and then spiked with $H_2O_2$ as needed. A small stir bar was added and mixing began at a rate that allowed for good mixing, but disturbed the surface of the water minimally.

Initial samples were taken for absorption measurement (0.7 mL), EDC analysis (0.4 mL) and $H_2O_2$ analysis (0.1-0.2 mL), where applicable. The samples were then irradiated in the MP or LP quasi-collimated beam UV radiation setup for the proper times corresponding to the desired delivered UV fluence, and sub-sampled (0.4 mL) for EDC analysis at various fluence values. At no time were more than 5 fluence values utilized, so the volume of sample taken for sub-sampling would not significantly change the depth of sample. For example, a water initially having a depth of 2.83 cm, after initial samples and 5 fluence sub-samples, would now have a sample depth of 2.74 cm, corresponding to 97% of the initial depth. This error was considered acceptable, so no corrections were made to account for the change in depth in the fluence calculation.

## BIOANALYTICAL METHODS FOR EDCS DETECTION

### *In vitro* E-Screen Assays (cell proliferation assay)

Initial evaluations of estrogenic activity were performed using the E-screen. MCF-7 breast cancer cells were requested from Dr. Ana. Soto at School of Medicine at Tufts University. They were MCF-7- BOS (passage 122) cells, and passage 127-143 was used in the study. The E-screen assay was carried out according to methods provided by Dr. Soto's lab (personal communication). Briefly, estrogen dependent MCF-7 cells were cultivated and maintained in Dubecco's modified Eagles medium (DMEM) (ICN Biomedicals) supplemented with 5% defined fetal bovine serum (FBS) (HyClone) and 0.23% sodium bicarbonate under 5% $CO_2$/95% air at 37°C. Sub-confluent cells were trypsinized, and re-suspended with the culture medium. Cells were seeded in 24-well plates (ICN Biomedicals) at a density of 20-30 thousand cells/well and allowed to attach for 21-27 hours. After the attachment period, old media was aspirated and replaced by 0.5 mL of testing chemical solutions, which were diluted to desired concentrations with experimental media immediately prior to use. The experimental media contained phenol-red free DMEM supplemented with 5% charcoal dextran stripped FBS (CD-FBS), 4 mM L-glutamine (Gibco), Hepes and Penicillin (4 mL of 1.5M Hepes pH 7.6 and 1cc of 100,000 units of Penicillin per 400-425 mL bottle of medium). Each dilution was tested in quadruplicate. Negative controls consist of steroid-free experimental media only. Positive controls were run as a series of six dilutions of 17β-Estradiol ($E_2$) between $10^{-13}$ to $10^{-8}$ M. Assays were terminated during late exponential phase of cell proliferation (five days of incubation). Cell number was determined by measuring total cell protein contents using the sulforhodamine B (SRB) method (Skehan et al, 1990), where total cellular protein bound to SRB dye is directly proportional to cell number by measuring the extinction of SRB at 550 nm, determined by Perkin Elmer Bioassay reader HTS 7000. Quantitative evaluation of estrogenic potential was made relative to the hormone free control by determining proliferation effect (PE) (cell number maximum (chemicals)/ cell number maximum (negative control)).

The validation of the E-screen assay was performed by testing cell response to standard $E_2$. The maximum relative cell proliferation ratio (cell proliferation in steroid-free media was considered as 1) significantly increased between $10^{-8}$-$10^{-10}$ M of $E_2$ solutions, and decreased with subsequent dilutions. This dose and response relation revealed a similar trend as references for $E_2$ (Soto et al., 1995; Korner et al., 2000). However, our maximum ratio was 2.5 fold, which was not as high as the references (up to 4-7 fold). Also, it was difficult to maintain the quality control for performance of the E-screen analysis. Therefore, the Yeast Estrogen Screen (YES) assay was substituted as the *in vitro* assay for these studies.

### *In vitro* Recombinant Yeast Estrogen Screen (YES) Assay

YES assays measure the ability of a single- or mixture of compounds to initiate transcriptional activation of a recombinant estrogen receptor-reporter gene in the engineered yeast cells. A human estrogen receptor (hER) and estrogen response elements (ERE) coupled to lac reporter gene are integrated into yeasts (*Saccharomyces cerevisae*). Wild type yeasts do not express the estrogen receptor (ER) or estrogen responsive genes containing a functional ERE. In the engineered yeast the ERE is linked to a reporter gene lac-Z. The protein product of this gene is the enzyme β-galactosidase which is secreted into the medium. Induction of β-galactosidase occurs when ER ligands (such as EDCs) bind to ER causing it to interact with the ERE and initiate the expression of lac-Z gene (Routledge and Sumpter, 1996). Then β-galactosidase in the medium then metabolizes the chromogenic substrate ortho-nitrophenyl-beta-D-galactopyranoside (ONPG), resulting in a colorimetric change of the ONPG at 405 nm (Huggett et al., 2003). These engineered yeast cells were originally provided by Dr. Sumpter at Brunel University in U. K. The YES assay was performed according to Sapozhnikova et al (2004) with some modifications. The changes include the use of 96 deep well plates in place of 1 mL microtubes, shorter incubation times (from 6 days to 3 days) by increasing initial cell numbers and the use of $OD_{420}$ instead of $OD_{405}$ to increase the sensitivity of the assay.

The yeast cells were grown in 5 mL growth media (containing 0.7% Difco yeast nitrogen base (YNB) nutrient powder (Becton, Dickinso and Company, Spark, MP), 0.52% ammonium sulfate, 0.002% adenine sulfate, 2% dextrose and 0.5% casamino acids). After overnight incubation at 30°C, cells were diluted with assay medium (1% YNB, 0.5% peptone and 10% sucrose) to an optical density (O.D.) of 0.08 at 630 nm. 100 μL dilutions of EDCs solution including BPA (0-100 μM), NP (0-100 μM), $E_2$ (0.0001- 100 nM) and EE2 (0.0001- 100 nM) were added to 96 deep-well plates (1 mL capacity/well, VWR, West Chester, PA) containing 300 μL of diluted yeast solution (a total dilution factor of 4 was used to calculate the final dosing concentrations for each compound). Plates were covered with sterilized papers and incubated at 30°C for 3 days with gentle shaking. 17β-estradiol (0.0001- 100 nM) served as a positive control. After 3 days of incubation, 50 μL Z-buffer (60 mM $Na_2HPO_4$, 40 mM $NaH_2PO_4$, 10 mM KCl, 1 mM $MgSO_4$ and 50 mM β-mercaptoethanol) was added to lysis cells. 400 μL assay buffer (containing 0.1% ortho-nitrophenyl-beta-D-galactopyranoside (ONPG), 98.4% Z-buffer, 1.3% β-mercaptoethanol and 0.3 % SDS) was then added to each well. Plates were sealed with an adhesive seals membrane (USA Scientific Inc., Ocala, FL) and incubated at 30°C for 30 minutes. 200 μL of 1 M sodium carbonate was used to stop β-galactosidase reactions in the media. Plates were centrifuged at 3,000 rpm for 10 minutes. To determine O.D. at 405 nm and 630 nm, 100 μL of supernatant was taken from each well and then transferred to a new 96 well microtiter plate. All values were presented as $OD_{405}$ minus $OD_{630}$ to ensure appropriate normalization of the data.

Dose-response relationships of YES assays for 17β-$E_2$, 17α-$EE_2$, Bisphenol A (BPA) and 4-nonylphenol (NP) are shown in Figure 2.3. The order of the estrogenic activities for these four EDCs was $EE_2$>$E_2$>NP>BPA, which is consistent with reference values for this analysis (Routledge and Sumpter, 1996; Andersen, et al., 1999). $EC_{50}$ (concentration to induce half of maximum response) each compound was calculated and listed in Table 2.1. Estradiol equivalency factor (EEF) (a quotient of $EC_{50}$ for Estradiol and the test compound) was determined.

### *In vivo* Vitellogenin (VTG) Assays

The orange red strain of Japanese medaka fish (Oryzias latipes) was maintained in a flow-through system under standard aquaculture conditions at the Duke University Aquatics Laboratory. To reduce endogenous VTG levels in males due to free estrogen from females, six month-old mature males were separated from females for two weeks prior to experimentation. To develop dose- and time- response relationships for VTG protein expression, fish were exposed to final $E_2$ solutions between 100 and 0.1 μg $L^{-1}$. All solutions were made from concentrated stock solutions made in DMSO and added directly in embryo rearing medium (ERM) (Yamamoto, 1975) with vehicle solvent representing less than 0.1% of the final volume.

Exposure durations included 72 hr, 1 week and 2 weeks. Each exposure contained 12 fish placed into aerated 2 L beakers containing 1L of testing solutions. Solvent (DMSO)-spiked ERM water was used as a blank control. Fish were feed with brine shrimp (*Artemia sp.*) daily. Half the volume of each exposure solution was renewed every day to control nitrogen accumulation and maintain water quality and space conditions for fish rearing. At each time point, four fish per group were anesthetized. Blood was collected from the caudal vein via glass-capillary tubes, and the volume of each sample was measured. All blood samples were centrifuged at 10,000 *g* for 15 min at 4°C to collect serum and serum was combined with protease inhibitors (0.2 mg $mL^{-1}$) aprotinin (Sigma, St. Louirs, MO) and 0.5 mM phenylmethylsulfonylfluoride (PMSF) (Sigma)). All serum samples were stored at -80$^0$C until use. The VTG protein in each sample was quantified by Immonodetection (Western blotting) analysis or Enzyme-Linked Immunosorbent Assay (ELISA) (Biosense Laboratories AS, Bergen, Norway).

**Western Blotting Analysis of VTG Induction**

Total protein concentration in serum was measured using the Bradford method (Bio-Rad protein assay kits, Hercules, CA). 5 μg of total protein for each sample was loaded onto a 10% SDS-polyacrylamide gel. Electrophoresis and Western blotting analysis for VTG protein were performed as described by Merril et al., (1998). Briefly, separation of proteins was carried out at 15 mA/gel for 2.5 hours in Tris-glycine buffer (1.4% glycine, 0.3 % Tris base and 0.1% SDS at pH 8.3). A full range rainbow recombinant protein marker (Amersjam, RPN 800) was used to determine the molecular weight of interested proteins in SDS gels. Following electrophoretic separation, proteins were stained with 0.0025% Coomassie Brilliant Blue R 250. The SDS-gel image is shown in Figure 2.4-left. A duplicate gel was transferred to an Immobilon$^{TM}$-P transfer membrane (Millipore, Bedford, MA) by electroblotting in Tris-glycine buffer with 10 % methanol at 400 mA for 3 hr. Membranes were blocked (4˚C over night) in 5 % (w/v) non-fat powdered milk in Tris-buffered saline (15 mM Tris and 500 mM NaCl, pH 7.5; TBS). Membranes were then incubated (1 hr, room temperature) with a primary anti VTG antibody (1: 10,000 dilution) and washed in Tris-buffered saline/0.05% Tween 20 solution. Membranes were subsequently incubated (1 hr, room temperature) with alkaline phosphate conjugated anti-rabbit secondary antibody (1: 7,5000 dilution) (Promega, Madison, WI). Following a final wash with Tris-buffered saline/0.05% Tween 20, protein targets were visualized with CDP-Star chemiluminescent substrates (Roche Molecular Biochemicals).

The result of VTG protein induction following 72 hr of $E_2$ exposure is shown in Figure 2.4-right. The plasma proteins are separated according to their molecular weights after electrophoresis. Kashiwada et al. (2002) and Hamazaki et al. (1987) reported a molecular weight of 200 kDa for a major protein subunit of medaka VTG under denaturing conditions and of 420 kDa for the native form of vitellogenin. Under our laboratory conditions, VTG protein was recognized as a single band between the protein markers of 160kDa and 250 kDa (Figure 2.4, left) and no protein band was observed at the corresponding locations for control (CTL) and 0.1 μg $L^{-1}$ E2 samples. Following Western Blotting, only VTG protein bands appeared, demonstrating the high degree of specificity of our VTG antibody (Figure 2.4, right). The overall results for VTG induction under laboratory conditions is demonstrated in Figure 2.5a. To numerically quantify VTG expression, O.D. from each VTG band was digitized by ImageQuant version 5.1 (Molecular Dynamics, Sunnyvale, CA), and normalized by the value of a positive control. The positive control was 5 μg of concentrated vitellogenin protein, which was collected from fish ascites fluid after $E_2$ exposure. For example, O.D. of the positive control was determined as 1.0 and blank control resulted in no response. VTG immunoreactivity was not detectable in the blood from untreated males (-CTL) and males treated with 0.1 μg $L^{-1}$ $E_2$ for 72 hr. However, plasma VTG concentrations increased concordantly with increases in dose and exposure time. Results suggest that VTG concentration in the blood may reach saturation when the $E_2$ concentration was over 10 μg $L^{-1}$ or exposure time was longer than 1 week.

### Enzyme-Linked Immunosorbent Assay (ELISA) of VTG Induction

The medaka VTG ELISA kit was purchased from Biosense Laboratories AS (Bergen, Norway). The principle of ELISA is similar to that of Western blotting except the reaction is performed in a 96-well plate format. ELISA utilizes specific binding between antibodies and VTG protein to quantify VTG in samples from Japanese medaka (Nilsen et al, 2004; Nishi, et al., 2002). The wells of microplates have been pre-coated with a specific capture antibody that binds to VTG from both standards and samples added to the wells. A different VTG-specific detecting antibody is added to create a sandwich of VTG and antibody, which is detected with an enzyme-labelled Secondary antibody. The enzyme activity is determined by adding a substrate that gives a colored product, and the color intensity is directly proportional to the amount of VTG present (guide book of Biosense Laboratories for medaka VTG ELISA kit).

ELISA was performed according to the protocol provided by Biosense Laboratories. Briefly, serum samples were thawed on ice, and dilutions of standard and samples were prepared before performing the assay. 100 μL of dilution buffer was added to the non-specific binding (NSB) wells of the plates as a blank control. Also, 100 μL of diluted standards and serum samples were added to the remaining wells. The plates were sealed with membranes and incubated at room temperature for 1 hour, and the plates were washed 3 times with 300 μL washing buffer per well. Then, 100 μL of diluted detecting antibody (primary antibody) were added to all wells. Plates were incubated at room temperature for 1 hour for antibody-antigen reactions followed by 3 washes. The identical procedure was performed with secondary antibody reactions except 5 times of 300 μL washing buffer per well was added at the end of reaction period. Finally, 100 μL of substrate solution was added to all wells, and incubated in the dark at room temperature for 30 min. 50 μL of 2 M $H_2SO_4$ was added to all wells to stop the reaction, and absorbance at 492 nm was determined (Biosense Laboratories manual book). On each plate, NSB-corrected absorbance values for standard and sample dilutions were calculated by subtracting absorbance values of the NSB wells. Linear regression analysis and 4- parameter transformation of the standard calibration curve (Figure 2.6a) were determined to calculate VTG concentration in the diluted samples. The VTG concentration in each original sample was obtained by multiplying the VTG concentration in the diluted sample with the dilution factor. Figure 2.6b shows VTG induction of medaka male fish following 7 days of water exposure of ethynyl estradiol ($EE_2$) at 0.5, 1.0 and 10 μg $L^{-1}$.

### *In vivo* Larval Toxicity Test for BPA

6 well plates (Corning, NY) were used to hold newly hatched (4-day old) medaka for larval toxicity evaluation. Each well contained 10 mL of BPA dilutions between 0-100 μM. 20 larvae were assigned to each well. Plates were kept in an incubator at 25°C. All solutions were renewed daily. Larvae were fed daily with ground Otohime β diet (Nisshin Co, Tokyo, Japan).

Dead larvae were removed daily and larval mortality recorded. Acute or sub-acute larval mortality of testing solutions was determined after the exposure of 72 h and 7 days respectively. Results are shown in Figure 2.7.

### *In vivo* Embryo Developmental Toxicity Assays for BPA

Embryo toxicity assays were performed in a 96-well plate format to increase the sensitivity and statistical significance of the assay (Carney, et al., 2005). The assays were validated with BPA standards (0-100 μM). Medaka eggs were collected and synchronized up to 10 hrs post fertilization from medaka brood stock. Filaments attaching adjacent eggs were broken by gently rolling clusters between moistened fingertips. Eggs were pooled and placed in NaCl solution (20g $L^{-1}$) for 30 min to reduce risk of parasitic or fungal infection. Eggs were subsequently rinsed in DI water and placed in ERM for final staging under a Nikon dissecting microscope (Iwamatsu, 1994). Eggs at multi-cellular stages (about stage 10) were selected and then placed in 96-well plates (one /well) containing 200 μL of serially diluted BPA standards. Twelve embryos were used for each replicate for a total of 36 embryos/ treatment. A blank control of embryos was exposed in ERM alone as a temporal reference for normal embryonic development. The whole plate was sealed with a breathable membrane and incubated at 25°C under a 16 hour-light/8 hour-dark photoperiod. Exposure solutions were renewed daily and continued for a total of 7 days. At the end of the exposure, embryos were rinsed in ERM, transferred to clean plates with ERM and followed through hatching. Direct, non-invasive observation of individual embryos was conducted daily through day 21 of development.

Mortality and sub-lethal endpoints including pericardial and peritoneal edema, cardiac, optic or spinal deformities, and hatching success were observed. Development of larvae including swim bladder inflation and swimming activity were additionally observed. Embryo mortality was determined by examining movement, heartbeat and blood circulation at 1-3 days (acute toxicity) and 3-10 days (sub-acute toxicity). Hatching success was determined as the breakdown and exiting of larval fish from the egg chorion. If an embryo did not hatch by day 21, it was counted as non-hatched even though clearly still alive. Cardiac abnormalities included pooled blood in the yolk sac, head or tail, pericardial edema (a sac of fluid surrounding the heart region and a clear wedge between the head and yolk sac) or formation of tube heart. Optic malformation includes asymmetric eyes and small microphthalmic or degenerative eyes (e.g. lacking a lens or complete pigmentation). Hatchlings were considered normal if they swam vigorously, had normal gross morphology, and inflated swim bladders. Medaka hatchlings inflate swim bladders within 24 hr. Evaluation was continued through the first three days after hatching (Villalobos et al., 2000; González-Doncel et al., 2003).

Figure 2.8 illustrates the hatching success ratio of medaka embryos as a function of exposure to BPA standards. $EC_{50}$ of BPA was determined as 73.9 μM with 95% CI between 68.6 and 79.7 μM using the EPA Trimmed Spearman-Karber (TSK) program version 1.5. Figure 2.9 illustrates control and BPA exposed embryos and hatchlings during development. The embryo on the left is representative of a normally developing individual (control) just prior to hatch (Figure 2.9a). For exposed embryos, hemorrhaging was observed at 30 mg $L^{-1}$ BPA. The hemorrhaging appeared as a large pooled blood in the yolk sac ventral to oil droplet. This embryo also showed a severe pericardial edema, appearing as a clear wedge between the head and yolk sac. The heart of this embryo developed as a narrow, pulsating, undifferentiated tube with circulating colorless fluid rather than a differentiated, two-chambered organ with circulating red blood cells. Furthermore, this embryo had microphthalmic and degenerative eyes without lens or complete pigmentation. The hatchling (bottom in Figure 2.9b) exposed to BPA (30 mg $L^{-1}$) during development exhibited severe cardiac abnormalities and non-inflated swim bladder (arrowhead pointed). Absence of swim bladder inflation may result in erratic swim behavior, increased gill movement, loss of equilibrium, and hyperexcitation (González-Doncel, et al., 2005). The corresponding control hatchling had inflated swim bladder (arrowhead) and normal heart structure (top larva).

**Comparisons of $EC_{50}$/$LC_{50}$ for Selected Bioassays**

To develop the linear working range and dose response relationship, biological assays were validated using a calibrated standard curve made by assessments of YES, VTG, larval and embryo toxicity assays. Validation of all bioassays was performed in triplicate. To calculate effective concentrations at half maximum response ($EC_{50}$) of BPA in YES and VTG assays, the sigmoidal concentration-response curve of the BPA standard was fitted to the Hill equation, equation 2.1, using the software Prism (GraphPad, San Diego, USA).

$$response = a + \frac{b - a}{1 + 10^{(\log EC50 - \log c)m}} \tag{2.1}$$

where a = baseline response (bottom), b = maximum response (top), c = concentration, m = Hill slope, and $EC_{50}$= half-maximal response. The lethal concentration at half maximum mortality ($LC_{50}$) of BPA in larval toxicity assays was determined by EPA Trimmed Spearman-Karber (TSK) program version 1.5. Table 2.1 lists $EC_{50}$ and $LC_{50}$ values for each assay and provides additional reference values identified from a literature review. These data are consistent with previously published values.

## Table 2.1
## Comparison of $EC_{50}/LC_{50}$ of four EDCs with selected bioassays

| | Environmental levels of EDCs [a] | YES assays (This study [f]/ references) | E-screen (This study / references) | VTG assays (Medaka); water exposure [f] | Larval toxicity assays (Medaka) (This study [f] / references) | Embryo toxicity assays (Medaka) (This study [f] / references) |
|---|---|---|---|---|---|---|
| Endpoints | Analytical methods | Reporter gene (β-galactosidas) expression | | 1 week-plasma VTG induction in male adults | 96 hr or 7 days mortality | Egg hatching |
| Unit | Max-Med conc. μM / ng $L^{-1}$ | μM | μM | μM | μM | μM |
| $E_2$ | (73-3.3)*E-05/ (200-9) | 3.44(3.20-3.71)*E-04/ 2.1E-04 [d] | 5.0*E-06/ 5.0*E-05 [c] | 1.1*E-03 (2 wk); 2.7*E-03 (1wk); 1.5*E-02 (72 hr)/ 6.4*E-04 (3wk;female zebrafish)[e] | - | 0.8/1.7 [b] |
| $EE_2$ | (28-2.5)*E-04/ (831-73) | 3.02 (2.85-3.09)*E-04/ 3.0E-04 [c] | 8.0*E-05 [c] | 2.1*E-5 (3wk;female zebrafish) [e] | - | - |
| BPA | (526-6.1)*E-03/ (12000-140) | 7.23 (6.46-8.07)/ 5.42-10 [d] | 0.4 [c] | 14.7 (11.6-19.4)/ >4.4 | 42.8 (40.5-45.2) (96h acute) 34.7 (31.8-37.8) (7day sub-acute) | 73.9(68.6-79.7) |
| NP | (182-3.6)*E-03/ (40000-800) | 2.2/ 8.0 [c] | 2.0 [c] | 1.1 (3wk;female zebrafish) [e] | - | 3.9 [b] |

[a] Max-Med concentration of EDCs in US river water by chemical analysis (Kolpin et al., 2002).
[b] Kashiwada et al., 2002
[c] Andersen et al., 1999
[d] Schultz et al., 1998
[e] Van den Belt et al., 2004
[f] Numbers in bracket represent the 95% confidence interval

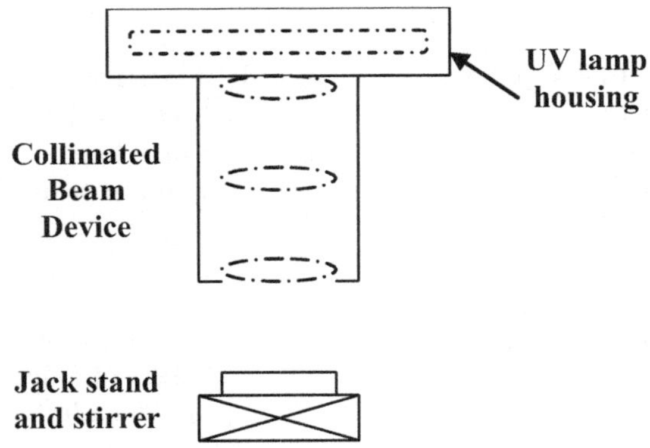


**Figure 2.1 Basic "Collimated Beam" device used in UV testing. Note circular aperatures within collimating "tube". Source: Reprinted in large part with permission from *Environ. Sci. Technol.*, October 15, 2004, *38*(20), 5476-5483. Copyright 2004 American Chemical Society.**

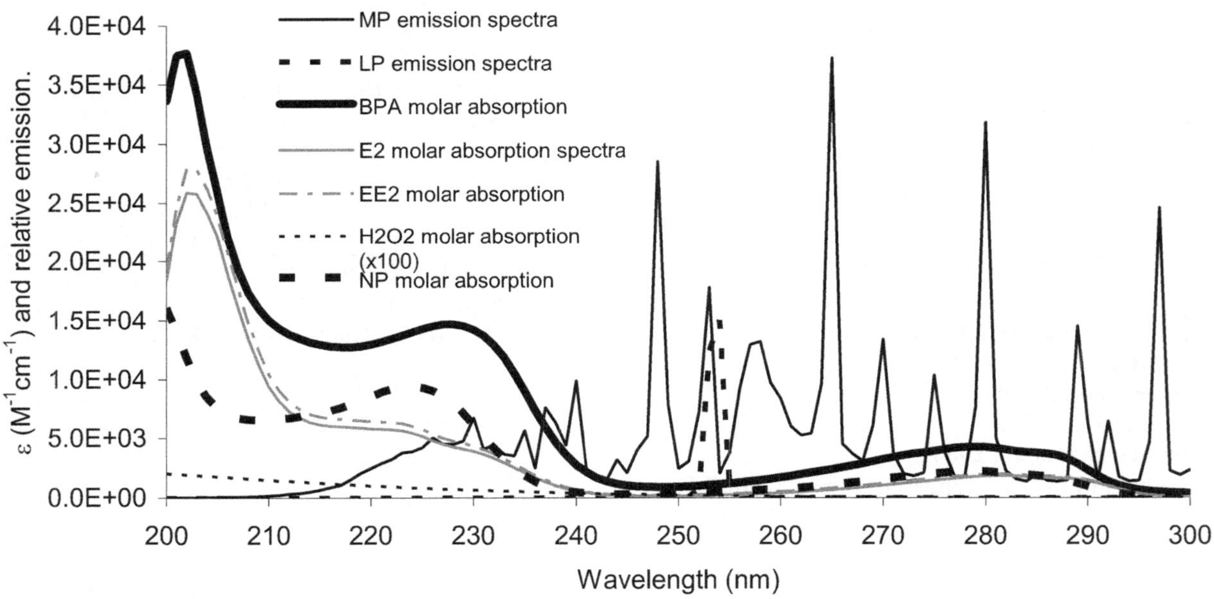


**Figure 2.2 Molar absorption spectrum of bisphenol-A (BPA), ethinyl estradiol (EE2), estradiol (E2), nonylphenol (NP), and $H_2O_2$, as well as the relative emission spectra of the UV lamps. Note, the molar absorption spectrum of $H_2O_2$ was multiplied by 100 to fit the scale. Source: Reprinted in large part with permission from *Environ. Sci. Technol.*, October 15, 2004, *38*(20), 5476-5483. Copyright 2004 American Chemical Society.**

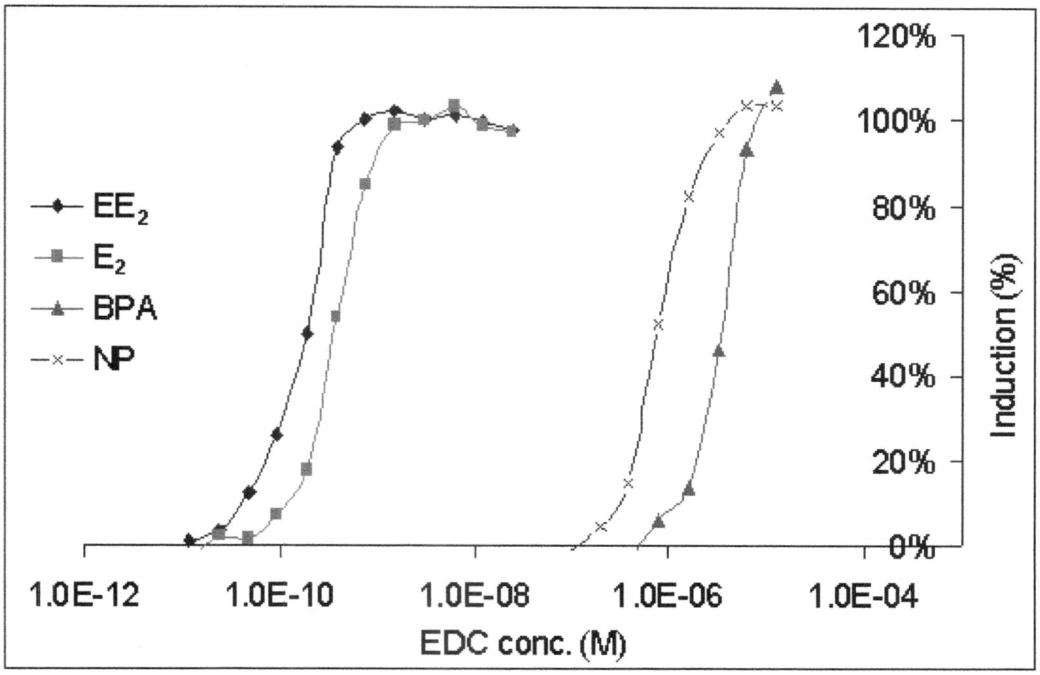


**Figure 2.3 Dose-response curve of YES assays for selected EDCs. Data are represented as the average of three replicates**

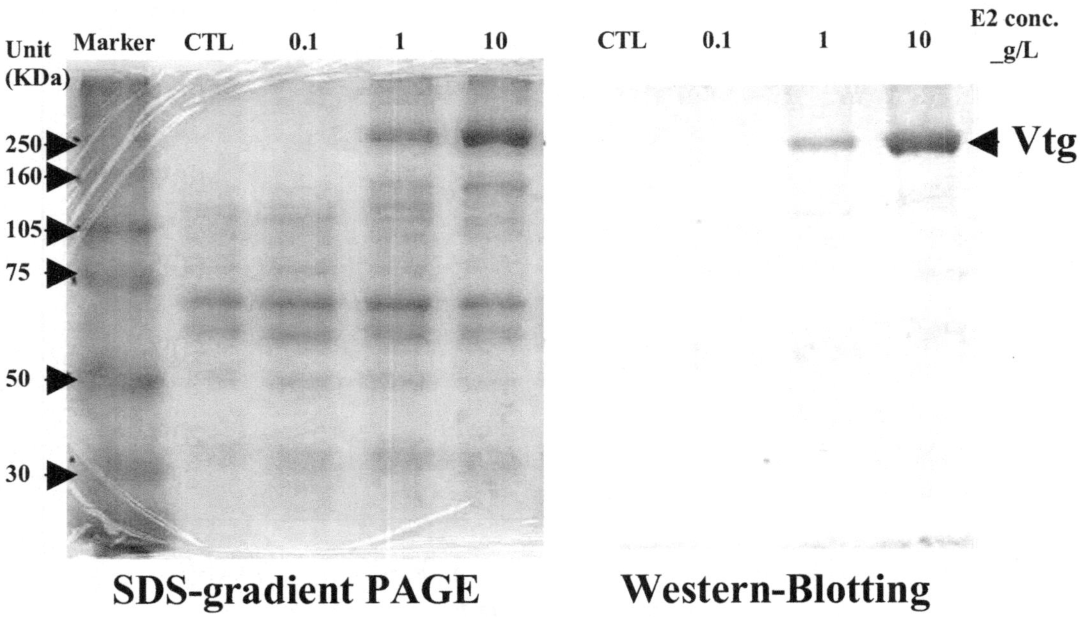


**Figure 2.4 Electrophoretic separation of plasma protein of Japanese medaka (left). Western blotting analysis of anti-VTG immune serum (right). Plasma protein of male medaka after 72-hr water exposure of $E_2$ was separated by SDS-PAGE and stained with Coomassie Brilliant Blue (left). A duplicate gel was electroblotted to a membrane and analyzed by Western blotting using anti-VTG immune serum (primary) and alkaline phosphate conjugated goat anti-rabbit (secondary) antibodies. The VTG protein (arrow pointed) was visualized with CDP-Star chemiluminescent substrates (right). The molecular weight of VTG protein in medaka plasma is approximately 210kDa (monomer).**

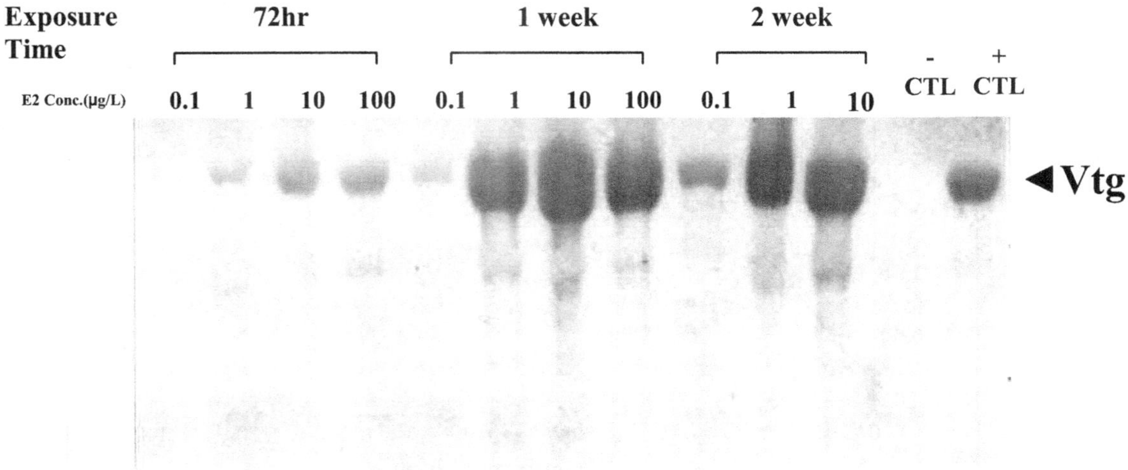


a)

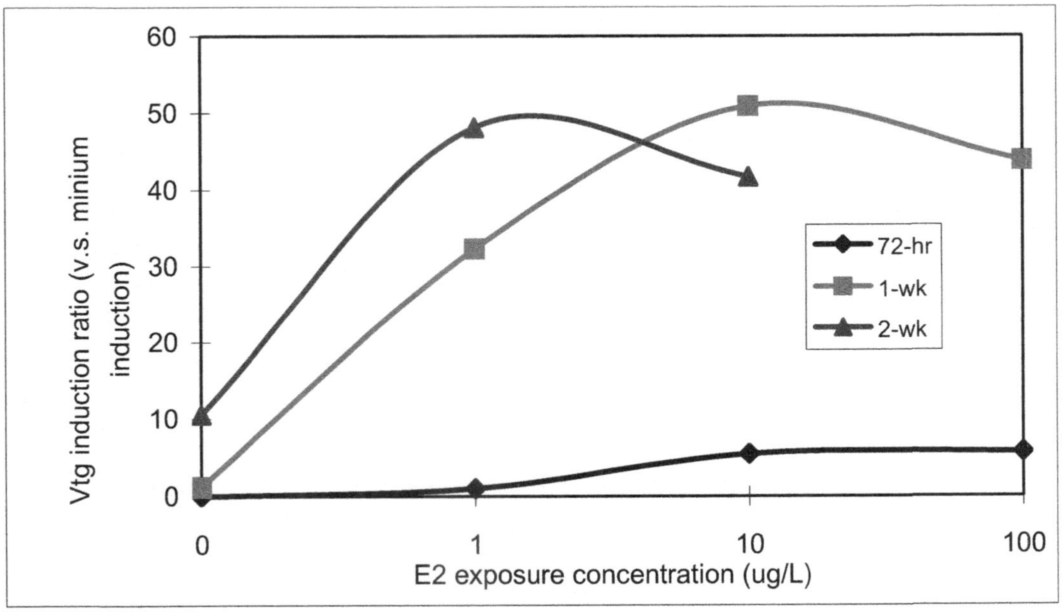


b)

**Figure 2.5 Dose- and time- dependent effects of E2 on vitellogenesis in male medaka. (a) Images of Western Blotting analysis; blood samples from E2-treated and untreated (-CTL) males were separated by a SDS gel and analyzed by Western blotting with antibody labeling after the gel was transferred to a nitrocellulose membrane. The band indicated by an arrow is VTG. The positive control (+CTL) is the VTG extracted from the ascites of males after 100 μg $L^{-1}$ of E2 exposure for 1 week. (b) Quantitative results of VTG inductions; the optical density of each band was digitized by Scion Image software and three dose and relative response curves were constructed compared with the minimum response (optical density of the minimum band at 1μg $L^{-1}$ -72hr exposure was determined as 1 to relatively quantify results)**

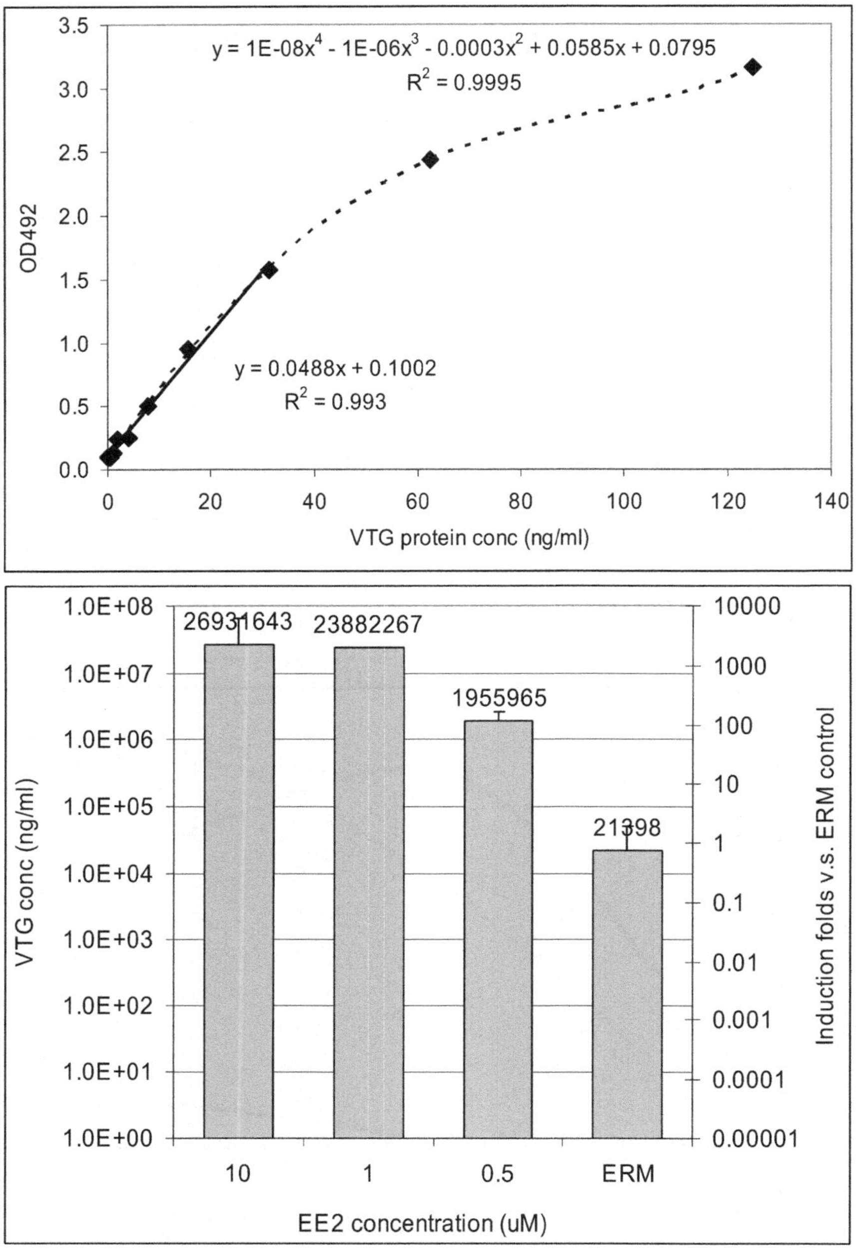


**Figure 2.6 a) Linear regression analysis and 4- parameter transformation of the standard calibration curve of VTG. b) VTG induction of medaka male fish following 7 days of water exposure of ethynyl estradiol (EE2) at 0.5, 1.0 and 10 µg $L^{-1}$. ERM is the blank control, showing endogenous VTG level of males after two weeks of separation from the females.**

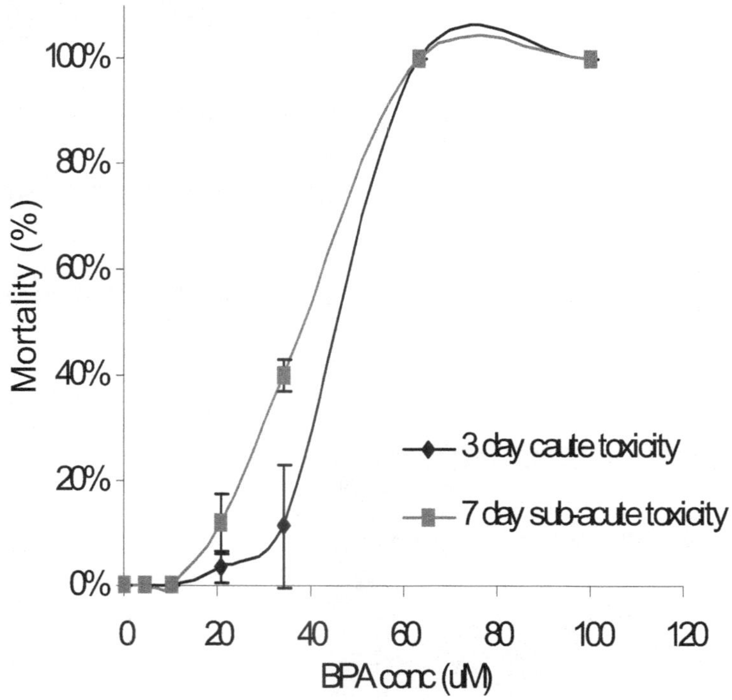


**Figure 2.7 Dose-response relationship between larval mortality and BPA exposure. Data represent as mean ± SD (n=3). 20 larvae/dose/replicate.**

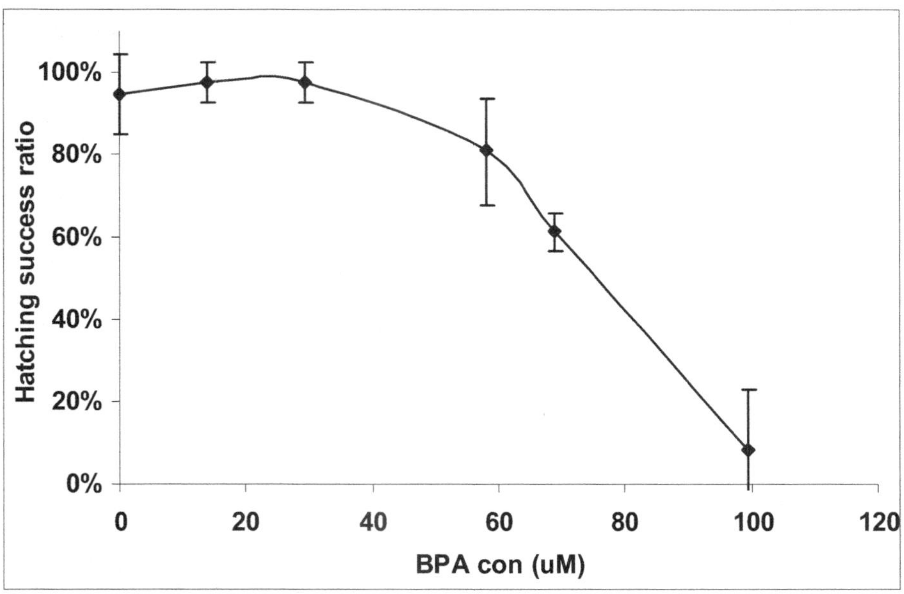


**Figure 2.8 Dose-responsive relationship between hatching success of embryo and BPA exposure. Data represent as mean ± SD (n=3). 12 embryos each replicate for a total of 36 embryos each concentration were used.**

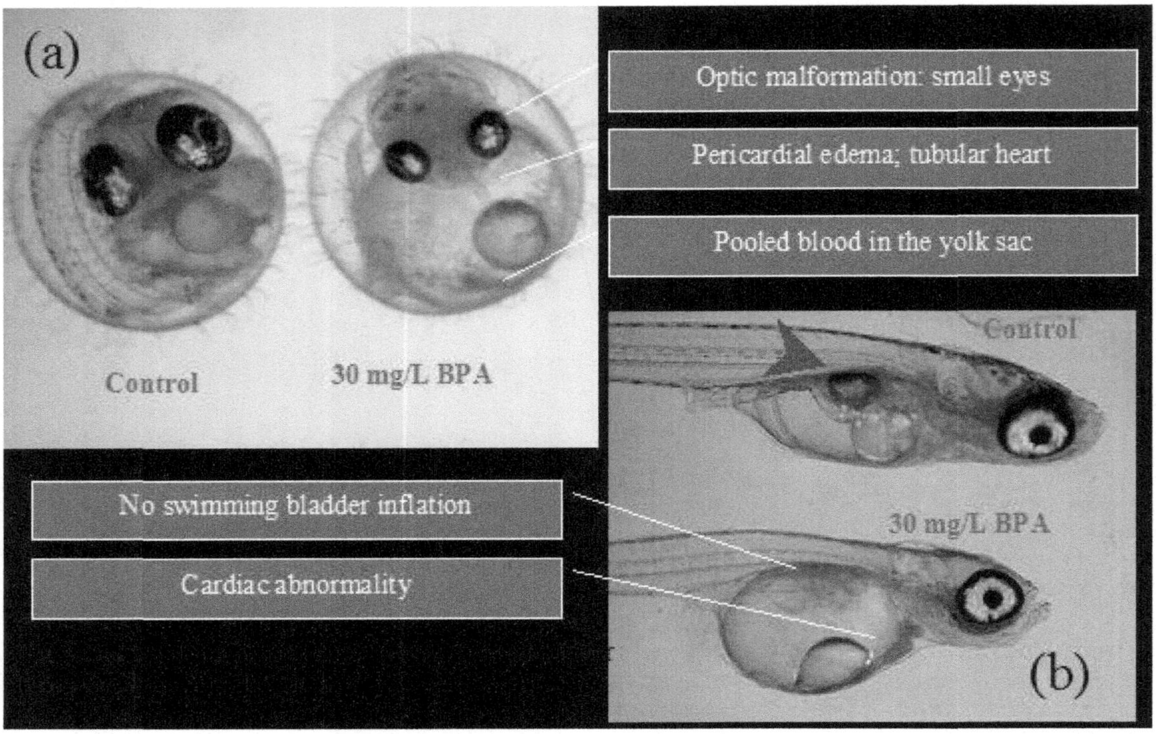


**Figure 2.9 Morphology of (a) embryos and (b) larvae of BPA aqueous exposure during developmental stages. (a) Embryo on the right showed severe hemorrhage after the exposure of 30 mg BPA $L^{-1}$; the embryo on the left is the corresponding control, which is a normally developing individual just before hatching. (b) The hatchling (bottom) is after the exposure to BPA (30 mg $L^{-1}$) during developmental stage, and the control is shown at the top.**

# CHAPTER 3
# FUNDAMENTALS OF UV PHOTOLYSIS AND OXIDATION OF EDCs

The effectiveness of UV based treatment is based on the capability of both direct photolysis and advanced oxidation processes. To evaluate how effective UV will be for treatment of contaminants, it is necessary to develop fundamental information regarding the quantum yield for the effectiveness of direct photolysis and hydroxyl radical rate constants for the effectiveness of advanced oxidation. This chapter presents research designed to measured these fundamental parameters and also model the UV based treatment processes of EDCs. This work was originally published in Environmental Science and Technology in 2004 (Rosenfeldt and Linden, 2004).

## EXPERIMENTAL

Each objective for this research was addressed by running a different set of experiments. These experimental sets are divided into the following five sections: direct photolysis, UV/$H_2O_2$ indirect photolysis, quantum yield determination, hydroxyl radical rate determination, and experimental model verification. A matrix of the experiments run to address each objective is presented below.

To calculate the quantum yield, direct photolysis experiments were run in laboratory water to find the pseudo-first order degradation constants under low and medium–pressure lamp conditions. This information, as well as irradiance data, was used to calculate the quantum yield.

Second order rate constants were determined for the EDCs based on a competing kinetics method (Zepp et al. 1987). A non-photoreactive compound, isopropyl alcohol (IPA, $k_{OH/IPA}$=1.9x$10^9$), was used as an OH radical scavenger to compete with the EDC for OH radicals. Low pressure irradiation experiments were performed in lab water samples containing an EDC, hydrogen peroxide (20mg/L), and Isopropyl alcohol (IPA). To avoid interference by IPA and EDC degradation byproducts, pseudo first order rate constants were determined by irradiating over several UV fluences of less than 150 mJ $cm^{-2}$. At this fluence, degradation of 4% of the initial IPA, 19% of the BPA, 20% of the EE2, and 26% of the E2 is expected.

Model validation experiments were performed in a model natural drinking water and natural Eno River water. The component concentrations in Table 3.1 (model natural drinking water), and water quality parameters including alkalinity, TOC, and pH (Eno River water) were entered into the UV/$H_2O_2$ model developed to predict the destruction of BPA. Irradiations were performed on the waters spiked with BPA and $H_2O_2$ and the data was compared against the model predictions.

### Direct UV

As Figure 2.2 illustrates, all the compounds absorb UV radiation in the 200-300nm range, with a significant absorption minima observed at approximately 250nm. Due to these observations, degradation of the compounds by exposure under monochromatic low-pressure UV lamps (emission at 254nm) was expected to be minimal, with significant improvements expected with the use of a polychromatic medium pressure lamp (emission throughout 200-300nm). As shown in Figure 3, a fluence of 1000 mJ $cm^{-2}$ elicits no EDC degradation of more than 5% under UV 254 from low-pressure lamps, meaning that after a fluence of 1000 mJ $cm^{-2}$, all of the compounds are present in at least 95% of their original concentration. However, 14.5%, 17.7%, and 21.6% (BPA, E2, and EE2, respectively) degradation occurs under the same fluence of polychromatic radiation from the medium pressure lamp.

The difference in degradation efficiency can be explained by the photochemistry caused by the absorption of additional UV wavelengths emitted by the MP lamp. The Grotthus-Draper law states that if, and only if, the radiation is absorbed by the molecule as a result of the interaction between the electromagnetic field associated with the molecule and that associated with the radiation, can the radiation be effective in producing photochemical changes (Wells, 1972). The molar absorption spectra of the EDCs (Figure 2.2) indicate that radiation emitted from the LP lamp (254 nm) has a low probability of being absorbed by the EDCs in the water, especially compared to the absorption of radiation between 220 to 230 nm or between 270 to 290 nm, covering wavelengths emitted by the MP UV lamp. As a reference point for EDC degradation via direct photolysis, typical UV disinfection applications utilize UV fluences between 20 and 120 mJ $cm^{-2}$.

## Quantum Yield

The quantum yield ($\Phi$) is defined as the moles of EDC transformed, divided by the moles of photons absorbed and can be calculated using equations 3.1 and 3.2 (Bolton, 1999).

$$\Phi(\lambda) = \frac{k_d^{'}}{\sum k_s(\lambda)} \qquad (3.1)$$

where

$$k_s(\lambda) = \frac{N_o(\lambda)\varepsilon(\lambda)[1-10^{-a(\lambda)z}]}{a(\lambda)z} \qquad (3.2)$$

In these equations, $k'_d$ is the pseudo-first order rate constant for direct photolysis of the EDC ($s^{-1}$), $k_s(\lambda)$ is the specific rate of light absorption by the EDC (Es $mol^{-1}$ $s^{-1}$) summed over all wavelengths for a polychromatic source, $N_o(\lambda)$ is the incident photon irradiance (Es $cm^{-2}$ $s^{-1}$), $a(\lambda)$ is the solution absorbance ($cm^{-1}$) and z is the depth of solution (cm).

Table 3.2 displays the quantum yields for each EDC in lab water using LP and MP radiation. Since these chemicals exhibit multiple absorption maxima, it is expected that the compound will have different quantum yields at different wavelengths. Quantum yield values for a compound are often constant over a small wavelength range, usually corresponding to the characteristic UV absorption bands of the chemical. However, calculating an overall quantum yield for LP and MP lamps is also very instructive to compare efficiencies for degradation between light sources. With this information, the efficacy of direct photolysis of each lamp can be determined, and compared in terms of energy, cost, and efficiency.

For direct photolysis to degrade a compound, photons need to be absorbed and the absorbed radiation has to be capable of degrading the contaminant. The efficiency of the absorbed radiation transforming the contaminant is described by the quantum yield, which depends on the probability of the excited state achieved by radiation absorption to evolve toward the products (Schwarzenbach et al., 1993). For the EDCs tested, the strong absorption bands observed in Figure 2.2 correspond to wavelengths that can cause the molecule to achieve a $\pi$-$\pi^*$ or n-$\pi^*$ excited electronic orbital state. When a molecule has been energized to these excited states, it is unstable and more likely to evolve towards products. Therefore, utilizing the radiation emitted by the MP lamp at these wavelengths created a greater possibility for the absorbed radiation to cause degradation, resulting in an increased quantum yield. The radiation emitted by the LP lamp is monochromatic at 254nm, corresponding to a wavelength that does not likely induce electron state transitions capable of transforming the EDCs, leading to low degradation and low quantum yields for this type of lamp.

Unlike the other EDCs tested, the quantum yield for LP UV treatment of NP is greater than that for MP UV treatment. This result seems counterintuitive and speaks to the chemical specific nature of the concept of quantum yield. The example of 4-nonylphenol serves to reinforce the fact that absorption spectra cannot be utilized alone to accurately predict the efficiency of degradation by absorbed UV radiation. The major structural difference between NP and the other EDCs tested is the nine carbon chain. This serves to make the molecule very hydrophobic, but may also minimize the effectiveness of absorbed UV energy at certain wavelengths to elicit chemical change. Additionally, LP UV seems comparatively more effective at degrading NP than the other EDCs. However, the standard deviation associated with the LP QY is large, speaking to several of the problems in solubility associated with the very hydrophobic molecule.

### UV/$H_2O_2$ Advanced Oxidation

To evaluate the destruction of the EDCs under oxidation conditions, hydrogen peroxide was added at low levels before exposure to UV radiation. As shown in Figure 2.1, significantly greater removal of each contaminant is achieved with the addition of approximately 15 mg/L hydrogen peroxide as compared to no addition.

The differences seen in destruction with the addition of hydrogen peroxide is due to the fact that the dominant mechanism of EDC destruction when $H_2O_2$ is added becomes hydroxyl radical mediated advanced oxidation. The AOP process involves the formation of the highly reactive OH radical species, which appears to quickly react with the EDC.

### Development of OH Radical Rate Constants

The second order rate constants for the reaction between the EDCs and OH radical was determined by utilizing a competing kinetics method with IPA based on previous work (Zepp et al. 1987). Equations 3.3a and 3.3b describe the theoretical base for the method, and a combination of these equations yields an expression that relates the observed pseudo first order rate constant for EDC degradation to the rate of OH radical formation, including the concentration of the main scavenging species present, IPA, $H_2O_2$, and the EDC.

$$k' = k_{EDC/OH}[OH]_{ss} \qquad (3.3a)$$

where

$$[OH]_{ss} = \frac{R_{OH}^{form}}{\sum k_{S,OH}[S]} \qquad (3.3b)$$

Substituting equation 3.3b into 3.3a and inverting the resulting equation yields equation 3.4, and by varying the concentration of IPA (0, 1e-5M, 2e-5M, 5e-5M, 7e-5M) as the scavenger in lab water, plots of the inverse of the EDC degradation pseudo-first order rate constants versus the concentration of IPA at constant $H_2O_2$ concentration and average UV fluence rate were created for each EDC.

$$\frac{1}{k'} = \frac{k_{IPA,OH}[IPA]_i}{k_{EDC,OH}R_{OH}^{form}} + \frac{[EDC]_i}{R_{OH}^{form}} + \frac{k_{H_2O_2,OH}[H_2O_2]_i}{k_{EDC,OH}R_{OH}^{form}} \qquad (3.4)$$

In these equations, k' is the pseudo-first order rate constant describing the destruction of the EDC by the $UV/H_2O_2$ process, $k_{EDC/OH}$ is the second order rate constant between the EDC and OH radical, $[OH]_{ss}$ is the steady state OH radical concentration, $R_{OH}^{form}$ is the rate of formation of OH radical by the $UV/H_2O_2$ mechanism, and $\Sigma k_{S,OH}[S]$ is the sum of the scavenger species present in the water. Figure 3.2 is a plot of the inverse pseudo first order rate constant for EE2 degradation as a function of initial IPA concentration, with slope $\left(m = \frac{k_{IPA,OH}}{k_{EDC,OH} R_{OH}^{form}}\right)$, and the y-intercept $\left(b = \frac{[EDC]_i}{R_{OH}^{form}} + \frac{k_{H_2O_2,OH}[H_2O_2]}{k_{EDC,OH} R_{OH}^{form}}\right)$. If the slope is solved for $R_{OH}^{form}$ and substituted into the y-intercept equation, this equation can be rearranged to solve for the time based second order rate constant $k_{EDC,OH}$ in terms of known initial concentrations and rate constants for BPA, EE2, and E2 as in equation 3.5.

$$k_{EDC,OH} = \frac{\left(\frac{bk_{IPA,OH}}{m} - k_{H_2O_2,OH}[H_2O_2]_i\right)}{[EDC]_i} \tag{3.5}$$

Table 3.3 displays the rate constants ($k_{EDC/OH}$) using the competition kinetics approach with Isopropyl Alcohol (IPA). The rate constant reported for EE2 compares well with an earlier reported rate constant of 9.8 +/- 1.2 $\times 10^9$ $M^{-1}s^{-1}$, within expected error (Huber et al, 2003). All of the rate constants indicate that the OH radical based decay of these EDCs is rapid and advanced oxidation is an effective treatment for degradation of these contaminants.

The traditional competition kinetics method, based on (Buxton et al. 1988), utilizing the reference compound para-chlorobenzoic acid (pCBA) was used to find the second order OH radical rate constant ($k_{OH}$) of NP. A solution of equimolar parts pCBA and NP (0.5 μM of each) in buffered water was prepared. To this solution, 25 ppm hydrogen peroxide was added, and the solution was irradiated, in triplicate, to a dose of 150 mJ $cm^{-2}$, with sub-sampling occurring at a dose of 30, 60, 100, and 150 mJ $cm^{-2}$.

**Degradation Modeling – $UV/H_2O_2$ AOP**

To understand the destruction of EDCs in water using $UV/H_2O_2$ AOP, a model capable of predicting the degradation of these species under different conditions was adapted (Sharpless and Linden, 2003) combining rate constants for both direct photolysis and indirect ($UV/H_2O_2$ AOP) photolysis. Briefly,

$$\text{Rate} = -\frac{d[C]}{dt} = (k_d' + k_i')[C] \tag{3.6}$$

Where $k'_d$ is the first order direct photolysis rate constant, and $k'_i$ is the pseudo-first order indirect photolysis rate constant. The first order rate constant for the direct photolysis ($k'_d$) of the contaminant C is calculated using the overall quantum yield Φ, and light intensity parameters via equation 3.7.

$$k'_d = \Phi\,(\Sigma k_s(\lambda)) \tag{3.7}$$

where $k_s(\lambda)$was previously was defined in Equation 3.2

To calculate the pseudo-first order rate constant for indirect photolysis, a steady state OH radical concentration assumption was made. The steady state OH radical concentration was calculated using equation 3.8, as the ratio of the formation of OH radicals to the destruction of the radicals.

$$OH_{SS} = \frac{\sum k_s(\lambda)\Phi_{OH}(\lambda)[H_2O_2]}{\sum_i k_{S,OH}[S]_i} \qquad (3.8)$$

where again $\Sigma k_s(\lambda)$ is as defined in equation 2, $\Phi_{OH}(\lambda)$ is the quantum yield for photolysis of $H_2O_2$ into 2 OH radicals (taken as 1 for the region below 300nm) (Baxendale and Wilson, 1957), and $\Sigma k_{S,OH}[S]$ is the sum of the product of the concentration of OH radical scavenging species [S] and the second order rate constants of the reaction between the scavenging species and OH radical.

The resulting indirect pseudo-first order rate constant is simply the product of the second order rate constant ($M^{-1}\ s^{-1}$) and the steady state OH radical concentration.

$$k'_i = k_{C/OH}[OH]_{ss} \qquad (3.9)$$

Once these pseudo first order rate constants are determined, destruction of the EDCs as a function of time is calculated by integrating equation 3.6 into the form

$$C_f = C_i e^{-(k'_d + k'_i)t} \qquad (3.10)$$

Model predictions and observed destruction data were compared for low and medium pressure UV combined with peroxide in several waters. The LP model was tested in model natural water, and raw, filtered Eno River water collected from near Penny's Bend Nature Preserve in Durham, North Carolina. The water was collected in a glass jar, and returned to the lab where it was immediately filtered through a 0.45 μm filter and refrigerated at 4°C for storage until use. The MP model was tested in lab water. Scavengers and corresponding OH radical rate constants used in the model included organic carbon ($2.5x10^4$ L $mg^{-1}\ s^{-1}$) (Larson and Zepp, 1988), $HCO_3^-$ ($8.5x10^6\ M^{-1}s^{-1}$), $CO_3^{2-}$ ($3.9x10^8\ M^{-1}s^{-1}$), and $H_2O_2$ ($2.7x10^7\ M^{-1}s^{-1}$) (Buxton et al., 1988).

**LP UV/$H_2O_2$ Destruction in Model Natural Water**

Figure 3.3 illustrates the predicted and experimental destruction of BPA in model natural water. Examining kinetics in model water is critical because it allows for observation of the destruction in a water environment with other scavenging species present, but because the water was prepared in the lab, the exact composition of the water was known. This facilitates accurate prediction of the steady state OH radical concentration as a function of water quality. Table 3.1 shows the exact concentration of each species in the model water used for this experiment.

Figure 3.3 represents data from an average of three nearly identical, but separate, model water runs. A model time-based rate constant was developed by averaging important photochemical parameters, and finding the averaged steady state OH radical concentration for the three runs. The bars surrounding the data points represent the high and low values for the three runs. The model accurately predicts the destruction at low fluence but as the fluence increases, the variability between data points increases. The model predicted time based rate constant of $1.73x10^{-3}$ is 2% greater than the averaged observed rate constant of $1.69x10^{-3}$.

### LP UV/$H_2O_2$ Destruction in Eno River Water

To examine how accurate the model was in predicting BPA destruction in a natural water system, filtered water from the Eno River in Durham, North Carolina was spiked with BPA and subjected to the UV treatments. Water quality parameters, including TOC, alkalinity, and pH were taken in an attempt to characterize the scavenger content of the natural water and are displayed in Figure 3.4 along with the results of the model as compared with the experimental data.

Predicted destruction, after initial agreement, tends to slightly overestimate the observed destruction. This may be due to the fact that the water quality parameters do not account for every scavenger present in the water, or that the NOM that makes up a portion of the total dissolved organic carbon (DOC), may be more active at scavenging OH radical than expected. DOC is the main scavenger in the system, and the scavenging contribution of DOC is lumped and described by a single rate constant, but depending on how much of the DOC is NOM vs. other organic carbon, the actual rate constant might be slightly different. Additionally, as the UV fluence increases, NOM is destroyed and the composition of the DOC is likely to change, but the current model does not account for this. The modeled OH radical concentration remains the same as if DOC was at its original concentration and composition, when in fact the OH radical concentration may be increasing as the contribution of the dominant scavenger is reduced. Even with these uncertainties, the difference between the predicted and observed pseudo first order rate constants is approximately 7%.

### MP UV/$H_2O_2$ Destruction in Lab Water

Figure 3.5 compares modeled results for BPA and EE2 destruction to actual data obtained for MP UV/$H_2O_2$ experiments in lab water. Visual examination of Figure 3.5 indicates good agreement between the predicted and observed destruction rates, and the predicted and observed pseudo-first order rate constants agreed to within 15% for both BPA and EE2.

Along with the difficulties of modeling the scavenging of any UV/$H_2O_2$ system, a few challenges specific to modeling a MP UV / $H_2O_2$ system may explain the small differences between the modeled and experimental results in Figure 3.5. Competition between the EDC and $H_2O_2$ for absorption of photons can lead to discrepancies between the energy allocated for direct photolysis compared to indirect photolysis, a situation that is magnified with MP UV, where direct photolysis is a significant contributor to the destruction.

Additionally, the byproducts formed by MP UV / $H_2O_2$ will theoretically be different than those formed by the essentially complete OH radical destruction of LP UV / $H_2O_2$. The mechanism of direct UV destruction often involves photo-hydrolysis, dimerization, or photoaddition. These are very different products than the suite expected by OH radical destruction, and could result in products that act as better OH radical scavengers than the EDCs being modeled. Further research regarding the products formed during these processes is currently ongoing, in an effort to better answer these and other questions.

The kinetic results presented in this study show that UV technologies can be an effective treatment scheme for degradation of several EDCs in water. A model, based on fundamental kinetic parameters has been effectively utilized to predict the destruction of these EDCs in various water matrices, including natural, surface water of known pH, alkalinity, and TOC. By utilizing this model, along with known kinetic parameters such as the OH radical rate constant and quantum yield for a compound, it is possible to predict the effectiveness of UV treatment processes for destroying the contaminant of concern. Several problems exist when "sizing up" bench scale data for use in real water treatment, including reactor geometry and hydraulics. By using "fluence-based" modeling along with the rate constants and quantum yields presented here, applications of these results to a reactor treatment scale can be made (Bolton and Stefan, 2002).

Of ultimate concern in any treatment process is an assessment of the treatment success in elimination of the action of the parent pollutant. Mineralization of the pollutant would require extended UV treatment times, and in many cases would not be practical. Thus, as with any oxidation process, breakdown products will be formed. The relative threats of these products, as compared to their parent compounds, must be assessed through toxicological testing. Byproduct speciation studies, as well as convenient estrogen screening methods are currently under examination in our laboratory to evaluate the effectiveness of the UV and UV/$H_2O_2$ treatment processes in reducing the threat posed to humans and the environment by these EDCs, and will be the subject of a future communication.

**Table 3.1**
**Composition of model natural water**

| pH | $[Na^+]$ (mg/L) | $[K^+]$ (mg/L) | $[Cl^-]$ (mg/L) | $[PO_4^{-2}]$ (mg/L) | Inorganic C, (as $CO_2$) (mg/L) | $[NO_3^-]$ (mg/L) | [NOM] (mg/L) | Alginic Acid (mg/L) |
|---|---|---|---|---|---|---|---|---|
| | | | | | | | | |
| 6.8 | 75.8 | 85.8 | 40.1 | 211.2 | 66 | 3 | 2.56 | 5.32 |

Source: Reprinted in large part with permission from *Environ. Sci. Technol.*, October 15, 2004, *38*(20), 5476-5483. Copyright 2004 American Chemical Society.

**Table 3.2**
**Quantum yield for each EDC evaluated**

| Compound | Lamp* | Quantum Yield (mol/Einstein) |
|---|---|---|
| Bisphenol – A (BPA) | LP | 0.0085 |
| Bisphenol – A (BPA) | MP | 0.019 |
| 17-α-Ethinyl Estradiol (EE2) | LP | 0.026 |
| 17-α-Ethinyl Estradiol (EE2) | MP | 0.061 |
| 17-β-Estradiol (E2) | LP | 0.043 |
| 17-β-Estradiol (E2) | MP | 0.10 |
| 4-Nonylphenol | LP | 0.12 +/- 0.04 |
| 4-Nonylphenol | MP | 0.05 +/- .0043 |

*Low pressure (LP) is at 254 nm; medium pressure (MP) is for the complete wavelength range absorbed by the compounds (200 to 300 nm)

Source: Reprinted in large part with permission from *Environ. Sci. Technol.*, October 15, 2004, *38*(20), 5476-5483. Copyright 2004 American Chemical Society.

**Table 3.3**
**Rate constants of EDCs with hydroxyl radical (OH)**

| EDC | $k^t_{EDC/OH}$ ($M^{-1}s^{-1}$) |
|---|---|
| Bisphenol – A (BPA) | 1.02 +/- 0.06 x $10^{10}$ |
| 17-α-Ethinyl Estradiol (EE2) | 1.08 +/- 0.23 x $10^{10}$ |
| 17-β-Estradiol (E2) | 1.41 +/- 0.33 x $10^{10}$ |
| 4-Nonylphenol * | 5.65 +/- 0.4 x $10^9$ |

* utilizing traditional competition kinetics based on Buxton et al. 1987
Source: Reprinted in large part with permission from *Environ. Sci. Technol.*, October 15, 2004, *38*(20), 5476-5483. Copyright 2004 American Chemical Society.

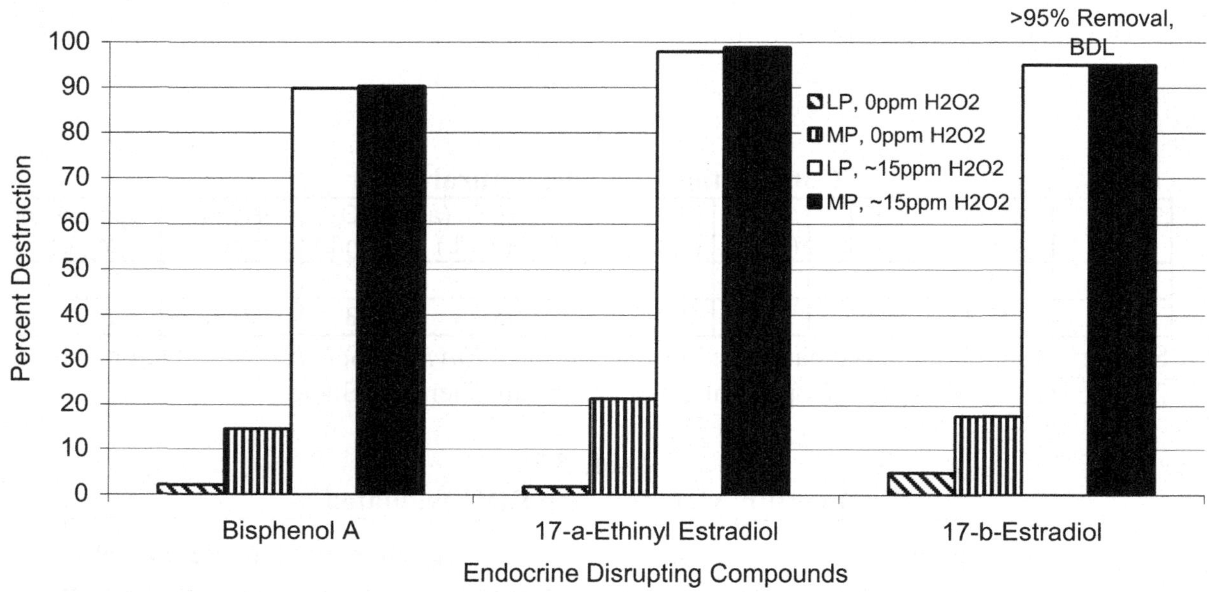


**Figure 3.1 Percent Destruction of BPA, EE2, and E2 following UV fluence of 1000 mJ cm$^{-2}$ via direct photolysis (0 mg/L $H_2O_2$) and UV/$H_2O_2$ AOP (~15 mg/L $H_2O_2$). The dark reaction between the EDCs and $H_2O_2$ produced no measurable EDC reduction at a time equivalent to a dose of 1000 mJ cm$^{-2}$. Experiment performed in lab water. BDL = below detection limit. Source: Reprinted in large part with permission from *Environ. Sci. Technol.*, October 15, 2004, *38*(20), 5476-5483. Copyright 2004 American Chemical Society.**

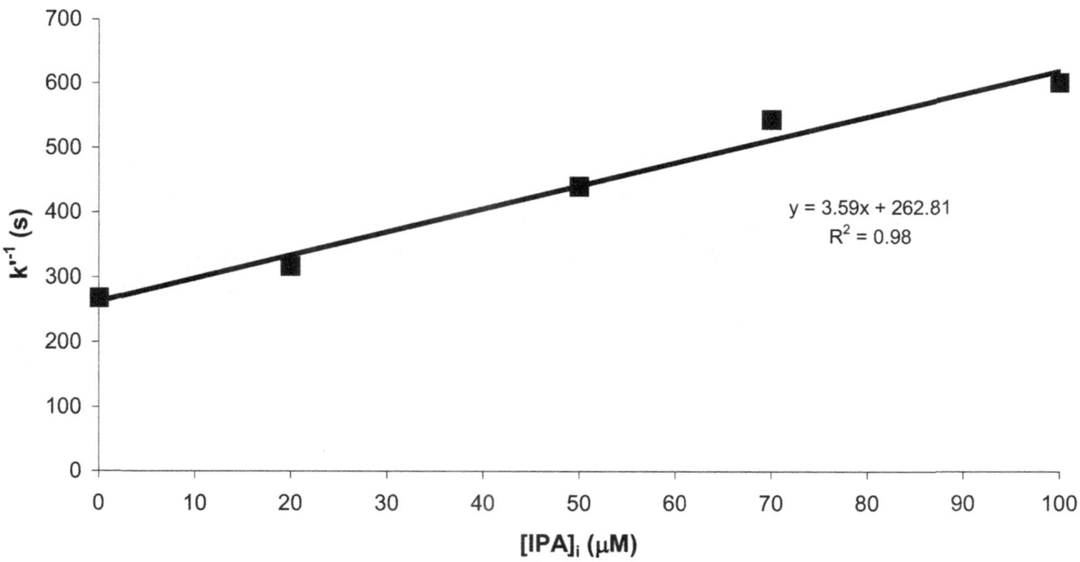


**Figure 3.2 Inverse pseudo-first order rate constant as a function of initial IPA concentration for use in calculating the second order rate constant of EE2 and OH. Initial hydrogen peroxide is 20 mg $L^{-1}$. Source: Reprinted in large part with permission from *Environ. Sci. Technol.*, October 15, 2004, *38*(20), 5476-5483. Copyright 2004 American Chemical Society.**

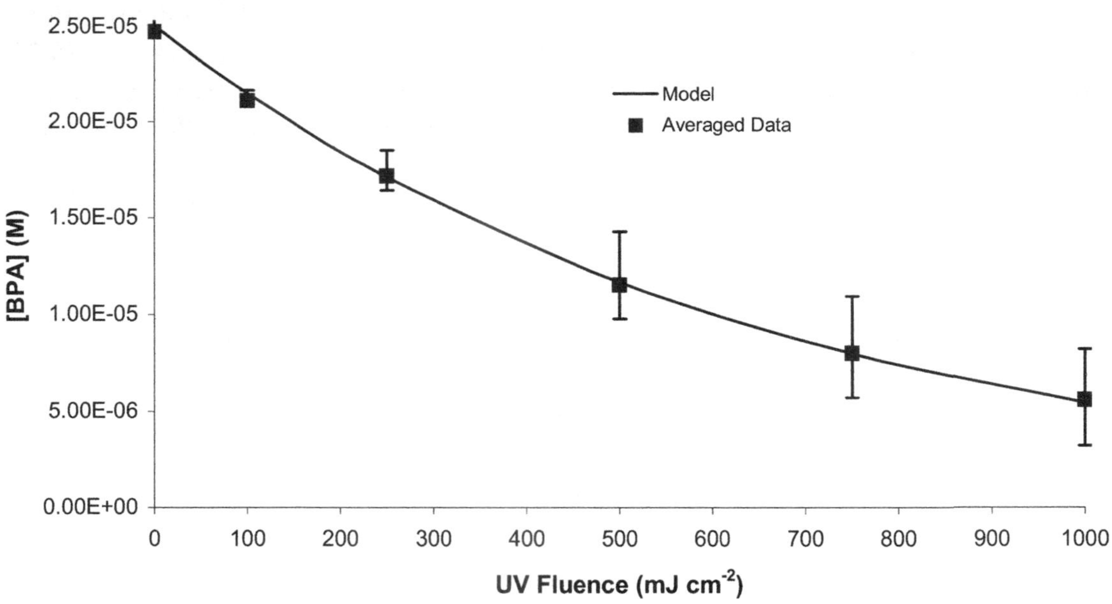


**Figure 3.3 Comparison of model predicted and experimental LP UV destruction of BPA dissolved in model natural water with 25 mg $L^{-1}$ $H_2O_2$. The solid lines represent the model prediction and the squares are the data points. Bars represent high and low values of the 3 replicates. Source: Reprinted in large part with permission from *Environ. Sci. Technol.*, October 15, 2004, *38*(20), 5476-5483. Copyright 2004 American Chemical Society.**

| H2O2 (ppm) | [BPA]I (M) | TOC (excluding BPA) (mg/L) | pH | Alkalinity (mg/L as CaCO3) |
|---|---|---|---|---|
| 14.21 | 2.33E-05 | 4.92 | 7.35 | 24.8 |
| 23.94 | 2.34E-05 | 4.9 | 7.35 | 25.8 |

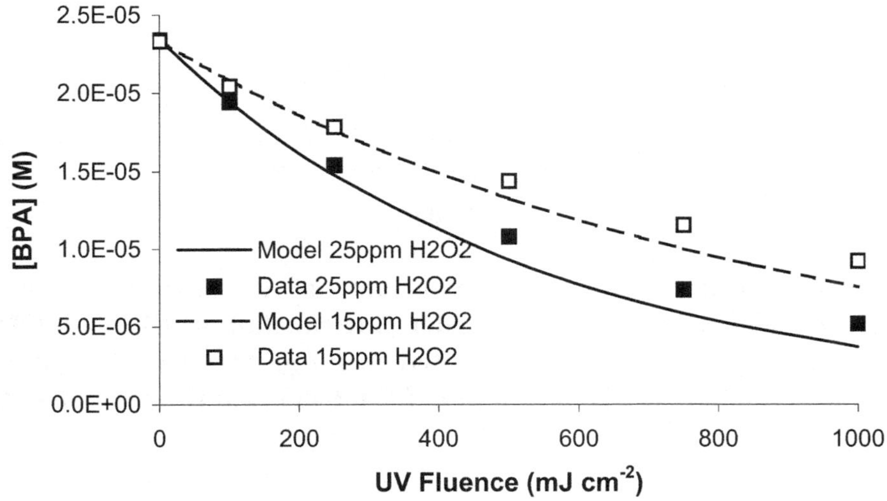


**Figure 3.4 Water quality parameters and comparison of LP UV/$H_2O_2$ model prediction and experimental data for BPA in Eno River water. Approximately 15 mg/L $H_2O_2$ and 25 mg/L $H_2O_2$ were used in the experiments. Source: Reprinted in large part with permission from *Environ. Sci. Technol.*, October 15, 2004, *38*(20), 5476-5483. Copyright 2004 American Chemical Society.**

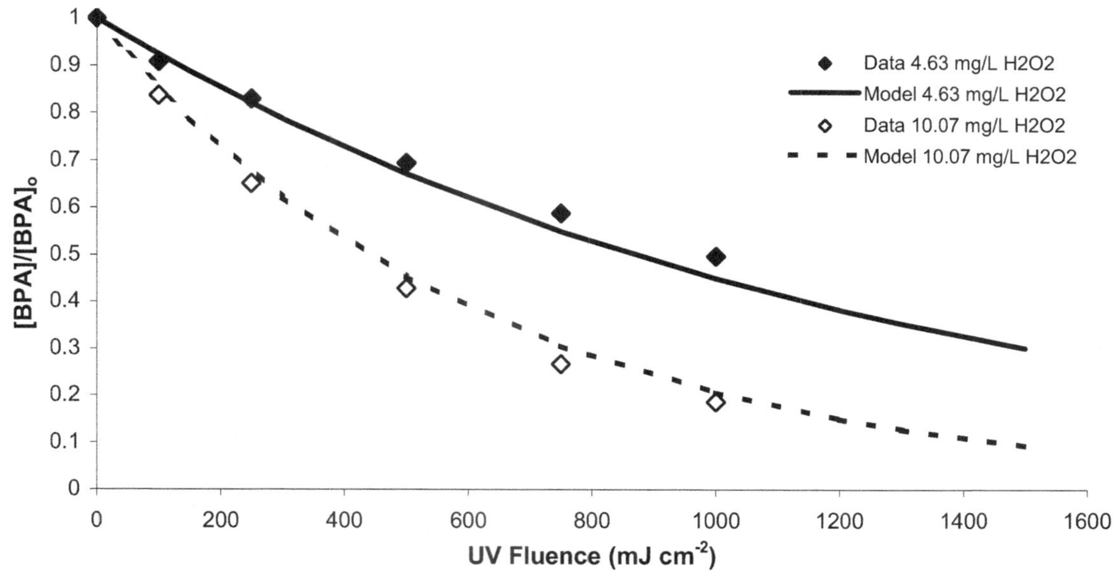


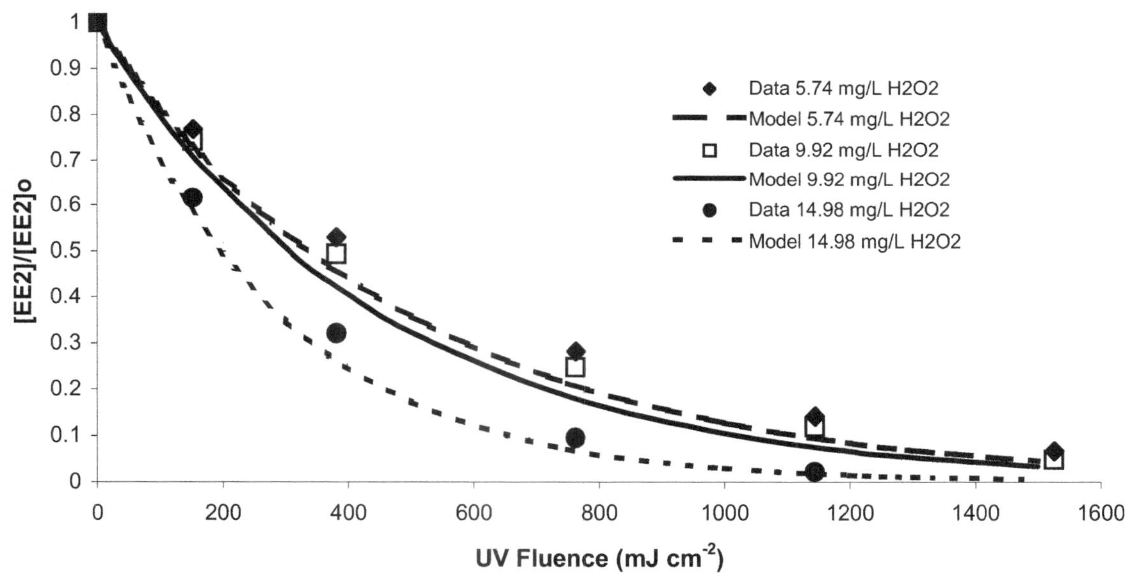


**Figure 3.5 Comparison between modeled results and actual degradation data for BPA (top) and EE2 (bottom) dissolved in lab water. The solid lines represent the model predicted values, and the symbols are experimental data. Source: Reprinted in large part with permission from *Environ. Sci. Technol.*, October 15, 2004, *38*(20), 5476-5483. Copyright 2004 American Chemical Society.**

# CHAPTER 4
# BIOANALYTICAL ANALYSIS OF UV TREATMENT OF BPA

## INTRODUCTION

Among numerous EDCs identified in environmental samples, bisphenol A (BPA) has received a great deal of attention from regulatory agencies and scientists due to a high level of production and widespread use in production of polycarbonates, epoxy resins, numerous plastic articles, and dental sealants (Staples et al., 1998). Studies have demonstrated that BPA affects the sex ratio, hatching, growth and mating behavior in both aquatic vertebrates and invertebrates (Segner et al., 2003). In terrestrial animals exposure of mice to BPA throughout embryonic/fetal life and during infancy increases oxidative stress, and is associated with underdevelopment of brain, kidney and testis (Kabuto et al., 2004). BPA has been detected at 0.14-12.0 $\mu g\ L^{-1}$ in U.S. river water (Kolpin et al., 2002), and up to 10 $mg\ L^{-1}$ in leachates from hazardous waste landfill sites in Japan (Yamamoto et al., 2001).

Due to the effectiveness of UV light for disinfecting drinking water, many water utilities are installing UV based systems. Such UV systems may also be effective for treatment of chemical contaminants. Rosenfeldt and Linden (2004) reported hydroxyl radical rate constants of $> 10^{10}\ M^{-1}\ s^{-1}$ for several EDCs including BPA, ethinyl estradiol and estradiol and demonstrated that the $UV/H_2O_2$ advanced oxidation process (AOP) is effective for degrading between 80% and 99% of these EDCs at a UV fluence of 1000 $mJ\ cm^{-2}$ in a simulated drinking water. Full mineralization of EDCs however, requires extended UV treatment and in many cases is not feasible option given the dynamics of the treatment train. Consequently, multiple unknown transformation products are formed at lower fluencies. While reduction in parent compound is a specific goal, it is additionally necessary to examine the potential biological activity of treated solutions that may contain byproducts with putative estrogenic activities. Therefore, the development and application of standard bio-analytical methods to thoroughly examine the biological activity/toxicity of the treated water, is important assessment tool.

Several *in vitro* bioassays have been widely applied to evaluate estrogenic activity of environmental samples including surface waters, wastewaters or treated water (Gray et al., 2002; Gomes et al., 2003). In a recent study, the recombinant yeast estrogen screen (YES) system was used to demonstrate a reduction in estrogenic activity of BPA and ethinyl estradiol following AOP technologies including $TiO_2$ photocatalyst and ozonation (Hu et al., 2002; Huber et al., 2004). Although *in vitro* assays provide fast and relatively inexpensive methods for quantification of aqueous estrogenic activity, these cellular systems fail to reflect the complexity of *in vivo* interactions due to differences of toxicokinetic and toxicodynamic parameters such as adsorption, distribution, metabolism and elimination. In addition, *in vitro* assays might over or under estimate the relative activity of EDC mixtures due to additive, synergistic or antagonistic interactions of EDC mixtures (Gray et al., 2002; Gomes et al., 2003).

Given several putative mechanisms of action for EDCs (e.g., estrogen receptor agonists/antagonists, and non- estrogenic receptor mediated actions), *in vivo* assessment in animal models is necessary to confirm *in vitro* estrogenic potency (O'Connor et al., 2002; Gomes et al., 2003). Japanese medaka fish have been extensively used as a model organism in reproductive toxicity testing. Their small size, relatively short generation times, asynchronous spawning, and high sensitivity to hormone administration at all life stages make them useful models for assessment of endocrine modulating activity (Ishikawa, 2000). The *in vivo* induction of the yolk protein, Vitellogenin (VTG), has been suggested as a highly responsive biomarker for the measurement of estrogenic potency of chemicals and effluents (Sumpter and Jobling, 1995).

The objective was to investigate direct photolysis with low- pressure (LP) Hg UV lamps and medium- pressure (MP) Hg UV lamps and UV + hydrogen peroxide ($H_2O_2$) advanced oxidation process (AOP) to assess the degradation of a prototypic endocrine disrupter, Bisphenol A (BPA), in laboratory water. Removal rates of BPA and formation of degradation products were determined by high performance liquid chromatography (HPLC) analysis. Changes in estrogenic activity were evaluated using both *in vitro* Yeast estrogen screen (YES) and *in vivo* Vitellogenin (VTG) assays with Japanese medaka fish (*Oryzias latipes*). The toxicity of BPA-treated water following UV/ $H_2O_2$ AOP was assessed with medaka larvae and embryos. The effectiveness of the MP-UV process, as compared to the LP-UV process, with and without $H_2O_2$ addition, is compared via biological and analytical assessments.

## MATERIALS AND METHODS

### Standards and Reagents

BPA was purchased from Aldrich Chemicals (Milwaukee, WI). BPA solutions for assessments of dose-responsive relationships of YES and VTG assays were prepared in Milli-Q deionized water or embryo rearing medium (ERM). Addition of a 10 mM BPA stock was dissolved in dimethyl sulfoxide (DMSO), and diluted with ERM for larval assays. The final solvent concentration in the medium did not exceed 0.1% (v/v). BPA stocks for UV irradiation were made in Milli-Q deionized water as 60 μM (13.7 mg $L^{-1}$) without addition of any carrier solvent. Stocks and diluted solutions were stored in amber vials at 4 $^0$C until use. All additional chemicals/reagents used during the experiments were analytical grade.

### Chemical Analysis of BPA and Degradation Products

Analytical equipment and methods are described in detail by Rosenfeldt and Linden (2004). Briefly, aqueous concentrations of BPA and the formation of degradation products were determined by high performance liquid chromatography (HPLC) (Varian, Inc., Palo Alto, CA) equipped with a C-18 (7.5 x150 mm) reversed phase Varian Prostar column and a photodiode array detector, set at 273-277 nm and 226-230 nm. The detection limit for BPA under the experimental conditions was ~0.1 μM for an injection volume of 50 μL. Following UV and UV/AOP treatments, BPA concentration (C) was normalized by the estimate of initial BPA concentration ($C_0$) as the remaining ratio of parent compounds ($C/C_0$).

### Experimental Design

#### *UV irradiation experiment*

The UV or UV/$H_2O_2$ AOP exposure was performed in a bench-scale collimated beam UV apparatus where UV lamps were suspended horizontally over an irradiation stage, with a stir plate and Petri dish containing the test solution. The UV apparatus consisted of four 15 W low pressure (LP) Hg lamps (ozone-free, General Electric #G15T8), which emit monochromatic radiation at 253.7 nm. Also, a 1 kW medium-pressure (MP) UV lamp (Hanovia Co., Union NJ) used in this study was housed in a benchtop collimated beam-type instrument (Calgon Carbon Corp., Pittsburgh, PA). The UV fluence (mJ $cm^{-2}$) was determined as the product of the average irradiance and exposure time based on calculations outlined in Bolton and Linden (2003).

***Direct photolysis and UV/$H_2O_2$ AOP with YES and chemical analyses***

120 mL BPA stock (60 μM) was placed in 70 × 50 mm flat bottom glass Petri dish and spiked with 0-50 ppm $H_2O_2$ (from a 30% stock solution) respectively. The solution was then exposed to UV irradiation with both LP and MP UV lamps at fluence ranging from 100-5000 mJ $cm^{-2}$ at ambient temperature. 0.6 mL of solution was sub-sampled following each UV exposure for YES and HPLC analyses. 0.2 ppm catalase (Sigma) was added to quench the residual $H_2O_2$, as measured by the Ghormley method (Klassen, 1994) to ensure the complete removal of residual $H_2O_2$. From each sample, 0.2 mL was used for HPLC analysis, and 0.3 mL of each sample was used for the YES assay. Aliquots of all samples were taken prior to treatment to establish base line values for HPLC, YES and absorption spectra. An $H_2O_2$ control consisting of 120 mL of 60 μM BPA and spiked with 50 ppm $H_2O_2$ was kept in dark for the duration of the study. A blank control of 120 mL of Milli-Q water spiked with 50 ppm $H_2O_2$ only was exposed under the same UV/AOP conditions as all test samples. All samples were subsequently stored in amber vials for YES assays.

The YES assay was performed according to the method by Routledge and Sumpter (1996) with significant modifications. Briefly, 100 μL dilutions of BPA standards (0-100 μM) and post treated solutions were added to 96 deep-well plates (1 mL capacity/well, VWR, West Chester, PA) containing 300 μL of diluted yeast solution at an optical density (OD) of 0.08 at 630 nm. In this format, a dilution factor of 4 was necessary to calculate the final concentrations of BPA (0-25 μM) in each well. Plates were covered and incubated at 30 °C for 3 d with gentle shaking. 17β-estradiol (0.0001-10 nM) served as a positive control. Following incubation, 400 μL of assay buffer (containing ortho-nitrophenyl-beta-D-galactopyranoside substrate and lysis buffer) was then added to each well. Plates were sealed with an adhesive seals membrane (USA Scientific Inc., Ocala, FL) and incubated at 30 $^{0}$C for 20 mins. 200 μL of 1 M $Na_2CO_3$ was used to stop β-galactosidase reactions in the media. Plates were centrifuged at 3,000 rpm for 10 mins. To determine OD at 405 nm and 630 nm, 100 μL of supernatant was taken from each well and then transferred to a new 96 well microtiter plate. All values were presented as $OD_{405}$ minus $OD_{630}$. The linear dose-response range of YES assays for BPA was between 1.5 and 15 μM (final concentration in the well). Prior to sample enrichments, detection limit for BPA under these experimental conditions was ~ 1.5 μM. All OD values from YES assays were converted to BPA equivalent concentration units via the transformation of a linear dose response regression curve of BPA standards. BPA equivalent concentration (C) was normalized to initial BPA equivalent concentrations ($C_0$) as $C/C_0$ and plotted with HPLC results. For OD values below the detection limit, samples were concentrated 100 fold by freeze-drying. Original (100 x concentrate) and subsequent serial dilutions of this concentrate were tested with YES assays. The recovery for freeze-dried samples was 101-111%.

***In vivo Vitellogenin (VTG) assessments of UV/$H_2O_2$ AOP***

To perform VTG assays for UV/$H_2O_2$ treated samples, 1 L of the BPA stock solution (60 μM), spiked with 10 or 50 ppm- $H_2O_2$ was placed in a 150 × 75 mm glass beaker and exposed to

LP and MP UV irradiation respectively between 0-5000 mJ $cm^{-2}$. A liter of deionized water spiked with 50 ppm-$H_2O_2$ following UV fluence of 5000 mJ $cm^{-2}$ was used as a blank control. After treatments, remaining $H_2O_2$ in the solution was quenched by addition of catalase (0.2 ppm) and the solution was stored at 4 C° for VTG assays. The orange red strain of Japanese medaka fish (*Oryzias latipes*) was maintained in a flow-through system under standard aquaculture conditions at the Duke University Aquatics Laboratory. To reduce endogenous VTG levels in males due to free estrogenic hormones in the aquarium, six month-old mature males were separated from females for two weeks prior to exposure. Four fish (six month old) were placed in a tank and dosed with 0.5 L of BPA standard solutions (0-50 μM BPA), UV or UV/AOP treated water for one week. Fish were fed with brine shrimp (*Artemia sp.*) daily. Half the volume of each exposure solution was renewed every day to control for nitrogen accumulation and maintain water quality. Aqueous BPA concentration in each exposure tank was monitored daily by HPLC.

After a 7 d exposure, two fish per group were sacrificed. Blood/serum were collected and combined with 0.2 mg $mL^{-1}$ aprotinin (Sigma, St. Louirs, MO) and 0.5 mM phenylmethylsulfonylfluoride (Sigma) to inhibit protease activities. Total protein concentration in serum was measured using the Bradford method (Bio-Rad protein assay kits, Hercules, CA). 10 μg of total protein for each sample was loaded onto a 10% sodium dodecyl sulfate (SDS)-polyacrylamide gel. Electrophoresis and Western blotting analysis for VTG protein were performed as described by Hames (1998) with some modifications. Briefly, separation of proteins was carried out at 15 mA/gel for 2.5 h in Tris-glycine buffer. Gels were transferred to Immobilon™-P transfer membrane (Millipore, Bedford, MA) by electroblotting in Tris-glycine buffer with 10% methanol at 400 mA for 3 h. Membranes were blocked (4 °C over night) in 5% (w/v) non-fat powdered milk in Tris-buffered saline. Membranes were then incubated (1 h, room temperature) with a primary anti-medaka VTG antibody (1:10,000 dilution) and washed in Tris-buffered saline/0.05% Tween 20 solution. Membranes were subsequently incubated (1 h, room temperature) with alkaline phosphate conjugated anti-rabbit secondary antibody (1:7,5000 dilution) (Promega, Madison, WI). Following two successive washes with Tris-buffered saline/0.05% Tween 20, protein targets were visualized with CDP-Star chemiluminescent substrates (Roche Molecular Biochemicals). To numerically quantify VTG protein expression, the optical density (OD) from each VTG band was digitized using ImageQuant version 5.1 (Molecular Dynamics, Sunnyvale, CA), and normalized by the value of a positive control. The positive control consist of 10 μg of concentrated vitellogenin protein, which was collected from fish ascites fluid following exposure to a high dose of estradiol ($E_2$). The OD of the positive control was determined as 1.0 and blank control resulted in no response.

### *Larval toxicity assessments of UV/$H_2O_2$ AOP*

Three replicates of UV/$H_2O_2$ AOP with either LP or MP UV lamps were performed for BPA solutions (60 μM) with addition of 50 ppm $H_2O_2$ at UV fluence between 0-5000 mJ $cm^{-2}$. Post treated solutions as well as a set of blank controls were collected. Six well plates (Corning, NY) were used to hold newly hatched (four day- old) medaka for larval toxicity assays. Each well contained 10 mL of treated solution and 20 larvae. Plates were kept in an incubator at 25 $^0$C. All solutions were renewed daily. Larvae were fed daily with ground *Otohime* β diet (Nisshin Co, Tokyo, Japan). Dead larvae were removed each day and larval mortality recorded. Acute or sub-acute larval mortality for each solution was determined after an exposure period of 72 h and 7 d respectively.

***Embryo toxicity assessments of UV/$H_2O_2$ AOP***

Three replicates of MP-UV/AOP were performed for BPA stocks with addition of 50-ppm $H_2O_2$ at UV fluence between 0-5000 mJ $cm^{-2}$. Post treated solutions as well as a set of blank controls were collected and used for embryo toxicity assays. Medaka eggs collected from a single day production by fish stocks were used. Eggs were pooled and placed in NaCl solution (20g/L) for 30 min to reduce risk of parasitic or fungal infection. Following rinsing in water, eggs were placed in ERM and staged under a dissecting microscope (Iwamatsu, 1994). Eggs at multi-cellular stages (about stage 10) were selected and then placed in 96-well plates (one /well) containing 200 μL of treated solutions. 12 embryos each replicate for a total of 36 embryos each treatment were performed. A blank control of embryos was exposed in ERM alone as a temporal reference for normal embryonic development. The whole plate was sealed with membrane and incubated at 25 $^0$C under a 16 hour-light/8 hour-dark photoperiod using a standard incandescent bulb. Exposure solutions were renewed daily and continued for a total of 7 days. At the end of the exposure, embryos were rinsed in ERM, transferred to clean plates with ERM and followed through hatching stages of development. Direct, non-invasive observation of individual embryos was conducted daily proceeding using dissecting microscope up to 21 days. Mortality and sub-lethal endpoints including pericardial and peritoneal edema, cardiac, optic or spinal deformities, and hatching success for embryos were observed.

**Data Analysis**

Validation of all bioassays was performed in triplicate. UV or UV/AOP treatment was performed in 2-3 parallel runs for HPLC and bioassays. Effective concentrations to half maximum response ($EC_{50}$) of BPA in YES and VTG assays were determined using Prism software (GraphPad, San Diego, USA) where data was fitted to the Hill equation. The lethal concentration to half maximum mortality ($LC_{50}$) of BPA in larval toxicity assays was determined by the EPA Trimmed Spearman-Karber program version 1.5.

Each UV treatment was performed in triplicate for both HPLC analysis and bioassays. Quantum yields for direct photolysis of BPA with LP and MP lamps were referred from Rosenfeldt and Linden (2004). Because direct photolysis and advanced oxidation of EDCs can be assumed as first or pseudo-first order reactions based on previous research (Rosenfeldt and Linden, 2004), the reaction rate is expressed as equation 4.1,

$$-\frac{d[C]}{dFluence} = k^{'}[C] \tag{4.1}$$

Here $C$ represents either BPA concentration for HPLC data, BPA equivalent concentration for YES data, or relative VTG induction concentration. $k$'is the overall rate constant, which represents either the direct rate constant for direct photolysis or a combination of the indirect and direct rate constants for UV based advanced oxidation. Upon rearranging and integrating (eq. 1), the equation is expressed as equation 4.2,

$$\ln\left(\frac{[C]}{[C]_0}\right) = k' \times Fluence \qquad (4.2)$$

A simple linear regression model for the statistical relationship between $\ln(\frac{C_0}{C})$ and UV fluence is expressed in equation 4.3,

$$\hat{\mu}\{\ln(\frac{C_0}{C}) | Fluence\} = \hat{\beta}_1 \times Fluence + \hat{\beta}_0 \qquad (4.3)$$

$\hat{\mu}\{\ln(\frac{C_0}{C}) | Fluence\}$ is the estimate of mean of $\ln(\frac{C_0}{C})$ as a function of UV fluence. $\hat{\beta}_1$ is the estimate of slope of the regression line, representing a fluence-based overall rate constant; $\hat{\beta}_0$ is the estimate of intercept of the line. 95% confidence interval (CI) for $\beta_1$ was computed with JMP IN software (SAS Institute Inc. NC, USA). The correlation coefficient ($r$) measured the linear relationship between $\ln(\frac{C_0}{C})$ and UV fluence ($r$=1 indicating a perfect linear relationship and $r$=0 indicating no linear relationship).

## RESULTS AND DISCUSSION

### Comparisons of Selected Bioassays

To develop the linear working range or dose-responsive relationships, biological assays were validated using a calibrated standard curve made by assessments of YES (0-25 μM BPA), VTG (0-50 μM BPA), or larval mortality (0-100 μM BPA) assay. The $EC_{50}$ value of BPA in the YES assay is 7.23 μM with a 95% confident interval (CI) between 6.46 and 8.07 μM. The $EC_{50}$ for the VTG assay is 14.7 μM with 95% CI between 11.6-19.4 μM. $LC_{50}$ of BPA for acute larval and sub-acute larval mortality is 42.8 (95% CI = 40.5-45.2) μM and 34.7 (95% CI = 31.8-37.8) μM respectively. $EC_{50}$ of egg hatching success in medaka embryos is 73.9 (95% CI = 68.6-79.7) μM. These data are consistent with previously published values for BPA estrogenicity (Schultz et al., 1998; Fang et al., 2000; Tabata et al., 2003).

### Results with Low Pressure UV lamps

#### *Removal of in vitro estrogenic activity compared to parent compound reduction*

Figure 4.1 illustrates the removal of total estrogenic activity (assessed by YES assays) as well as the decrease of parent BPA concentration (measured by HPLC) in solution following UV and UV/$H_2O_2$ AOP treatments. By using a BPA concentration of 60 μM, a direct analysis of BPA and BPA metabolites could be determined using HPLC and YES assay. This initial concentration was based on the calculated detection limits for both HPLC and the YES assay and enabled direct analytical assessment of individual samples without further concentration. A high concentration of 15 μM BPA equal to a one-quarter dilution of the 60 μM stock, gave a maximal response in the YES assay. In some instances following UV/AOP treatment (e.g. 25 ppm $H_2O_2$ at 5000 mJ $cm^{-2}$ or 50 ppm $H_2O_2$ at > 2000 mJ $cm^{-2}$), concentrations of BPA and estrogenic activity dropped below our detection limits. To further evaluate BPA degradation and biological activity, these samples were concentrated 2 orders of magnitude by freeze-drying. As illustrated in Figure 4.1a, direct LP-UV exposure without addition of $H_2O_2$ was ineffective for BPA degradation or reduction in estrogenic activity. With addition of $H_2O_2$ between 10-50 ppm, the removal efficiency increased sharply (Figures 4.1b-d). For example, 80% of parent BPA was degraded at 5000 mJ $cm^{-2}$ with addition of 10 ppm $H_2O_2$. Analysis of treated water via the YES assay resulted in a similar removal (78%) of estrogenic activity under this treatment. With addition of 25 ppm $H_2O_2$, > 97% of parent compound and estrogenic activity was removed at 5000 mJ $cm^{-2}$. When $H_2O_2$ was added up to 50 ppm, an equal removal rate (97%) was achieved at lower UV fluence (3000 mJ $cm^{-2}$). Greater than 99% removal of parent compound and estrogenic activity was achieved at 5000 mJ $cm^{-2}$. The BPA concentration in the dark control did not change significantly during experiments. It is important to note that the high concentrations of BPA resulted in significant scavenging of OH radicals away from the parent compound degradation as the daughter products remained in the mixture. Therefore, the high UV doses used are not necessarily relevant for UV treatment of BPA at environmental concentrations but are utilized here to illustrate the difference in parent compound decay as compared to decay of estrogenic activity.

Interestingly, $C/C_0$ of estrogenic activity was larger than the calculated disappearance of the BPA. The difference between the two values is assumed to represent the putative estrogenic activity of transformation products (shown as gray dashed lines). These data suggest that while the parent compound is transformed and the total estrogenic activity is decreasing, certain degradation byproducts are produced which retain estrogenic activity. These data may reflect activity of a single or possibly a mixture of estrogenic degradation product(s) that act in an additive or synergistic fashion with the remaining BPA parent.

***Formation/disappearance of degradation products***

HPLC analysis of BPA following photooxidation in the presence of 50 ppm $H_2O_2$ resulted in the production of several degradation by products. (Figure 4.2). Two distinct intermediate peaks appeared at retention times of 3.7 and 3.2 mins. These two peaks increased at first, then decreased with higher UV fluence. These data suggest a sequential degradation mechanism with increasing UV fluence. The HPLC peaks were grouped according to their retention time and chemical properties as Peak 4.3 min (parent compound), Peak 3.7 min (separate peak), Peak 3.2 min (separate peak), Peak 1.5-3.0 min (relatively hydrophilic fractions) and Peak 1.2 min (the rest of compounds dissolved in the carrier solvent/water). The appearance/disappearance of BPA degradation products under four treatments is summarized in Figure 4.3. Peak 3.7 min was present with each treatment, and is shown to have a similar trend to the formation of putative estrogenic activity illustrated in Figure 4.1 (gray dashed lines in the bottom each figure).

Previous reports utilizing $TiO_2$-photocatalyst or Fenton reagent indicated enhanced transformation of BPA in water via hydroxyl radical oxidation. Analyses of LC/MC and LC/MS/MS for the transformation products suggested the formation of common products such as *para* substituted phenols (Ohko et al., 2001; Hu et al., 2002; Fukahori et al., 2003). ρ-Hydroxyacetopohenone and ρ-Isopropenylphenol were detected with slight estrogenic activity using the YES assays (Ike et al., 2002). Since UV/$H_2O_2$ AOP, $TiO_2$-photocatalysis and Fenton reaction have been proposed to decompose BPA via oxidation by hydroxyl radicals, similar degradation products or intermediates of BPA may be formed. Therefore, our predictions on UV and UV/$H_2O_2$ transformation of BPA suggest the possibility of generating *para* substituted phenolic intermediates, which may exhibit estrogenic activity. We are currently in the process of identifying the chemical composition of these metabolites. HPLC analysis (Figure 4.3) demonstrates that full mineralization of BPA was not achieved even at the strongest AOP condition (50 ppm- $H_2O_2$/5000 mJ $cm^{-2}$). The sum of all peaks from 1.5 to 3.0 min increased with UV fluence, and reached a maximum between 2000 and 3000 mJ $cm^{-2}$ with addition of 50 ppm $H_2O_2$. This group of peaks was found to consist of several products when BPA treated solutions were concentrated (Figure 4.2e).

***In vivo* VTG assessments following LP-UV/AOP treatments**

*In vivo* estrogenic activity of BPA solution following UV irradiation with addition of either 10- or 50- ppm $H_2O_2$ was assessed by evaluating VTG protein synthesis. The semi-quantitative analysis of immuno-blotting for VTG induction is shown in Figure 4.4. Medaka VTG protein in plasma has a molecular weight of 420 kDa in the native form, and a major protein subunit of 210 kDa under denaturing conditions (Kashiwada et al., 2002). Results show that our antibody is highly specific to medaka VTG protein as a single band between the protein markers of 160 kDa and 250 kDa; in addition, no VTG is observed at the corresponding locations of blank control. Similar to YES/HPLC analyses, Figure 4.4a demonstrates VTG induction significantly decreased with UV fluence with addition of 50 ppm $H_2O_2$ in the solution. Under this treatment, a UV fluence of > 1000 mJ $cm^{-2}$ resulted in 40% reduction on VTG response in the exposed medaka although > 80% of parent compound was degraded. Furthermore, fish exposed to treated sample following 2000 mJ $cm^{-2}$ resulted in no detectable VTG response, suggesting complete removal of estrogenic activity under this condition. In contrast, LP-UV/10-ppm $H_2O_2$ treatment demonstrates a slower reduction of VTG protein expression with increasing UV fluence compared with parent compound reductions (Figure 4.4b). For example, at 5000 mJ $cm^{-2}$, the post-treated solution resulted in only a 40% VTG reduction and 80% removal of parent BPA compound. This result did not match our YES data for the LP-UV/10-ppm $H_2O_2$, in which ~80% of both estrogenic activity and parent compound was removed at 5000 mJ $cm^{-2}$. These results demonstrate a significant discrepancy between our *in vitro* and *in vivo* analyses. For both treatment conditions (UV/AOP + 10 and 50 ppm $H_2O_2$), the estrogenic activity (measured via vitellogenin responses) was much stronger than that observed with YES assays.

## Results with Medium Pressure UV lamps

### *Removal of in vitro estrogenic activity and parent compound*

Figure 4.5 illustrates the removal of YES measured estrogenic activity and the decrease of BPA (measured by HPLC) solution following direct UV irradiation with MP lamps and MP-UV+$H_2O_2$ AOP treatment. As hypothesized, direct UV exposure with MP-UV lamps (without addition of $H_2O_2$) was ineffective for reduction of BPA estrogenic activity (~4% removal) up to 5000 mJ $cm^{-2}$, while 22% of initial BPA parent compound was degraded at this fluence. With addition of $H_2O_2$ between 10-50 ppm, the removal efficiency increased sharply (Figures 4.5(b)-(d)). For example, 30% and 69% of parent BPA was degraded at 1000 and 3000 mJ $cm^{-2}$ respectively with addition of 10 ppm $H_2O_2$. Analysis of treated water via the YES assay resulted in only 20% and 57% removal of estrogenic activity at 1000 and 3000 mJ $cm^{-2}$ respectively. When spiking with 25 ppm $H_2O_2$ in the solution, >91% of parent compound and 84% of estrogenic activity was removed at 3000 mJ $cm^{-2}$. The required fluence to achieve this level of treatment was even lower when $H_2O_2$ was added at a concentration of 50 ppm. Interestingly, removal of estrogenic activity was slower than the disappearance of the BPA under direct photolysis and UV+10 ppm $H_2O_2$ AOP, suggesting production of estrogenic degradation product(s) and additive or synergistic response between remaining BPA parent and any estrogenic degradation products produced under these conditions.

### *Degradation products formed by MP-UV or MP-UV/$H_2O_2$ AOPs*

The transformation of BPA into degradation products under MP-UV treatments is illustrated in Figure 4.6. Figure 4.6(d) demonstrates that even under the strongest treatment condition (50 ppm $H_2O_2$ + 5000 mJ $cm^{-2}$ irradiation), full mineralization of BPA was not achieved. The components and trends of BPA degradation products formed by MP-UV or MP-UV/ $H_2O_2$ were similar to those formed by LP-UV or LP-UV/ $H_2O_2$ except for the UV+25 ppm $H_2O_2$ treatment. Interestingly, at this treatment level, the MP-UV+ $H_2O_2$ AOP outperformed the LP based AOP determined by reduction in estrogenic activity. Although the identity of these degradation products is not yet complete, previous reports using hydroxyl radical oxidation have identified a number of degradation products such as phenol or hydroquinone. These products are found to exhibit specific biological toxicities (Atkinson and Roy, 1995; DeCaprio, 1999), which in sufficient concentrations can be toxic to aquatic life.

### *Removal of in vivo estrogenic activity*

*In vivo* estrogenic activity of BPA following MP- UV irradiation combined with either 10- or 50-ppm $H_2O_2$ was assessed via VTG assays. The semi-quantitative analysis of immuno-blotting for VTG induction is illustrated in Figure 4.7. Similar to YES/HPLC analyses, Figure 4.7 (a) demonstrates VTG induction significantly decreased with increasing UV fluence in the presence of 50 ppm $H_2O_2$. Under this treatment, a UV fluence of 1000 mJ $cm^{-2}$ resulted in 52.6% reduction of VTG response in the exposed medaka although 80% of the parent compound had been degraded. When exposed to samples treated at 3000 mJ $cm^{-2}$ with 50 ppm $H_2O_2$, the fish showed no detectable VTG response, indicating a complete loss of estrogenic activity. In contrast, MP-UV/10-ppm $H_2O_2$ treatment demonstrated a less rapid decrease in VTG protein expression with increasing UV fluence, in comparison with parent compound reductions (Figure 4.7(b)). For example, at 3000 mJ $cm^{-2}$, the post-treated solution reduced 43% of VTG induction compared with 64% removal of BPA parent compound. The difference between *in vitro* and *in vivo* analysis may be due to differing phamacokinetic and mechanistic processes of these two assays.

**Comparisons of removal efficiency between LP- and MP- UV systems**

To further compare the removal efficiency of BPA or estrogenic activity between LP- and MP- UV systems, fluence-based rate constants are given in Table 4.1 for analytical (HPLC) and bioanalytical (YES and VTG) data.

***Direct photolysis***

Table 4.1 shows direct UV irradiation with MP lamp resulted in a slow first order reaction ($k$=1.04 $cmJ^{-1}$; $r$=0.96) for degrading BPA parent compound, while LP-UV direct photolysis did not show significant first order reduction ($k$=1.01 $cmJ^{-1}$; $r$=0.31). Quantum yields for BPA under MP and LP irradiation were determined to be 0.0084 and 0.0057 mol $Es^{-1}$ respectively. Figure 2.2 illustrates a substantial overlap of the emission from MP lamp with absorption band of BPA (centered at 225 and 278 nm), while BPA has minimum absorption near 254 nm (monochromatic emission of LP lamps). This very weak absorption, combined with a low quantum yield helps to explain why the LP lamp emission was not capable of transforming BPA, and the resulting slow degradation. The removal rate constant for YES estrogenic activity with MP lamps was also slightly higher than that observed with LP lamps. The linearity of the relationship between $\ln(\frac{C_0}{C})$ and UV fluence for YES data was not as good as for the HPLC data, indicating the removal kinetics for the estrogenic activity via direct photolysis may not be first order.

***$UV/H_2O_2$ AOP***

Table 4.1 illustrates that removal rate constants of either BPA parent compound or estrogenic activity significantly increased with the increase of $H_2O_2$ concentrations in the solution for both lamps. In addition, for both UV lamps the removal rates of *in vivo* VTG estrogenic activities were statistically lower than those observed *in vitro* (YES data). However, there was no statistical evidence supporting any difference between MP or LP UV/AOPs with the addition of $H_2O_2$ between 10-50 ppm. Linearity of the relationships between $\ln(\frac{C_0}{C})$ and UV fluence for VTG data was not as strong as found for YES data, indicating the removal mechanism via *in vivo* estrogenic activity was more complicated than that predicted *in vitro*.

Overall, the comparison analysis demonstrates that removal rates of *in vivo* estrogenic activities are statistically lower than those observed *in vitro*. Perhaps explanations for discrepancy between *in vitro* and *in vivo* data reflect differences in pharmacokinetic and pharmacodynamic properties between these two bioanalytical models. For example, 1) YES assay incorporates human estrogen receptor ERα which may have different binding affinities and kinetics compared to the medaka estrogen ERα. 2) Most organisms express 2-3 different ER forms: our *in vivo* data may reflect differential binding of BPA and or metabolites to one or multiple forms of these ER's. 3) There may be both ER and non-ER based mechanisms for VTG induction *in vivo*, as well as 4) differences in adsorption, distribution, metabolism, and excretion (ADME). One consistent criticism of the YES assay is reduced sensitivity due to reduced uptake of EDCs across the yeast cell wall. Comparatively, fish readily accumulate BPA and other lipophilic substances.

**Larval and Embryonic Toxicity of UV/AOP Treated Water of BPA**

***Decrease in larval mortality***

Acute (72 h) and sub-acute (7 d) toxicity of BPA solutions was evaluated following different irradiation times in the presence of 50 ppm $H_2O_2$. While initial BPA solution resulted in 43% acute or 58% sub-acute toxicity, larval survival rates in LP-UV/$H_2O_2$ AOP treated solution significantly increased (Figure 4.8 a), and suggest that toxic intermediates are not produced under these treatment conditions. However, following 2000 and 3000 mJ $cm^{-2}$ treatment, sub-acute exposure resulted in a slight decrease in larval survivability. It is highly possible that the slight increase in larval mortality is due to the transient production of transformation products such as phenol or hydroquinone, which exhibit specific biological toxicities (Atkinson and Roy, 1995). Replicate BPA solutions were irradiated with MP-lamps under same conditions (+ 50 ppm- $H_2O_2$), and similar results (Figure 4.8 b) were found as observed with LP-UV AOPs.

***Changes in Embryo toxicity***

Figure 4.9a shows the hatching success ratio in medaka embryos under the exposure of MP-UV/AOP treated solutions (for 60 μM BPA in addition of 10 and 50 ppm $H_2O_2$) as function of UV fluence. Results show hatching success increased up to 97 % when 3000 mJ $cm^{-2}$ was used for both treatments. However, the ratio decreased to 41.7 % (arrowhead) when embryos were exposed BPA solution in the presence of 50 ppm $H_2O_2$ after 5000 mJ $cm^{-2}$ of UV irradiation (Figure 4.9b). In contrast, embryos completely hatched while exposed to the same solution with 10 ppm $H_2O_2$ under same UV fluence (Figure 4.9a). The pH of treated solutions decreased with fluence for both treatments, especially for the treated solution after 50 ppm- $H_2O_2$ + 5000 mJ $cm^{-2}$ treatment (pH= 4.3) (Figure 4.9b). A low pH of treated solution resulting in embryonic toxicity was found with LP-UV lamps for the same BPA solution in the presence of 50 ppm- $H_2O_2$ following 5000 mJ $cm^{-2}$ of irradiation (Figure 4.9a).

When morphology of embryos under the treatment condition of 50ppm- $H_2O_2$ at 5000 mJ/$cm^2$ was further examined, 92% (33 out of 36) of embryos were found to possess some granular precipitates inside the chorion (Figure 4.10 a & b). Of the embryos with this change, 58% (21 out of 36) died without hatching in 7 days, and 34% appeared to live normally (inserted figure in Figure 4.10 b). In addition, only 50% of resultant hatchlings had inflated swim bladders (Figure 4.10 c).

Possible explanations for embryonic acute/sub- toxicity in the treated solution (following UV/50ppm $H_2O_2$ at 5000 mJ $cm^{-2}$) included low pH effects due to the formation of acidic products, presence of toxic degradation products or the combination of these two factors. To test whether low pH of solution would cause embryonic lethality, embryos were exposed to ERM at different pH values (between pH=3-6). Results show 100 % of embryos hatched successfully (Figure 4.11 a) and survived (Figure 4.11 b) in the ERM solution of pH=7 following 7 day-exposure; however, this ratio significantly decreased with the decreasing pH, and no embryos survive at pH =3. In an additional study, the pH of UV/AOP-treated solutions (following UV/50ppm $H_2O_2$ at 5000 mJ $cm^{-2}$) were adjusted from pH=4.3 to pH=~7. This neutralized treated solution was then used for embryo exposure to observe whether toxicity of embryos would disappear. Figure 4.11 shows that not only hatching but also survival ratios in embryos increased when the pH of the treated solutions was raised up to pH=~7. According to these results, UV+ $H_2O_2$ AOP likely produced acidic degradation products that lowered the solution pH, resulting in embryonic lethality. Granular precipitates inside the chorion were still present, however, following adjustment to neutral pH. Embryos exhibiting this phenotype proceeded to hatch without significant alterations. Embryo assays were limited to seven days and concentrations tested provided an acute assessment of toxicity. While in some instances UV/AOP retained estrogenic activity, chronic toxicity assessments were not conducted. Prolonged exposure to treatment byproducts with and without estrogenic activity may result in alternate biological effects, however this remains to be determined.

## CONCLUSIONS

Our studies demonstrate that the biological assessment of residual estrogenic activity in the treated solution remains an ideal way to determine effective treatment endpoints. UV alone did not effectively degrade BPA, whereas UV/AOP under adequate $H_2O_2$ and UV fluence were highly effective for removing aqueous estrogenic activity to below detectable levels. However, due to the high initial concentrations of BPA used for study of estrogenic activity, the UV fluence levels utilized were 50 to 100 times higher than what is typically applied in disinfection practice and 5 to 10 times higher than standard UV/AOP applications. The study of UV-based oxidation treatment at environmentally relevant concentrations of EDCs is the topic of Chapter 5, where we demonstrate that these high UV doses are actually not required for effective treatment of EDCs at low concentrations typically found in the environment.

*In vitro* and *in vivo* results demonstrated UV-AOP based oxidation with either LP- or MP- UV lamps was effective for removing aqueous estrogenic activity to below detectable levels. Via comparisons of UV fluence-based removal rate constants for parent compound or estrogenic activity between the two UV lamps, it was demonstrated that direct photolysis with MP-UV lamp was marginally effective to remove parent compound and YES estrogenic activity compared to LP-UV lamps. No significant difference in removal rates was found between the two lamps under UV+ $H_2O_2$ AOP. Our data suggest that while the parent compound was being transformed and total estrogenic activity was decreasing following UV/AOP, formation of certain active metabolites may still occur. This phenomenon was demonstrated both *in vivo* and *in vitro*. Removal rates of *in vivo* estrogenic activity were significantly lower than those observed *in vitro*. These differences highlight the differential sensitivities of *in vitro* and *in vivo* bioanalytical assays and reemphasize the necessity to employ mechanistically diverse assays for assessment of EDC activity in treatment studies. A battery of bioassays combined with chemical analysis is essential to examine the treatment endpoints. A strategy that encompasses the evaluation of aqueous biological activity via *in vitro* and *in vivo* bioassays for treatments of EDCs should be applied as a method to critically assess the efficacy of other water and or wastewater treatment processes. The use of *in vitro* assays such as the YES should be considered for EDC monitoring in natural waters.

**Table 4.1.**
**Comparisons of removal rate constants (*k'*) of BPA or estrogenic activity following UV or UV/ $H_2O_2$ AOP with chemical or biological analyses.**

| UV lamps | MP-UV | | | | | | LP-UV [d] | | | | | |
|---|---|---|---|---|---|---|---|---|---|---|---|---|
| $H_2O_2$ conc | HPLC | | YES | | VTG | | HPLC | | YES | | VTG | |
| (ppm) | *k'* [a] | *r* | *k'* [a] | *r* | *k'* [a] | *r* | *k'* [a] | *r* | *k'* [a] | *r* | *k'* [a] | *r* |
| 0 | 1.04 | 0.96 | 1.02 | 0.68 | | | 1.01 | 0.31 | 1.00 | 0.03 | | |
| 10 | 1.48 | 1.00 | 1.27 | 0.95 | 1.16* | 0.70 | 1.40 | 0.99 | 1.34 | 0.95 | 1.11** | 0.81 |
| 25 | 2.13 | 1.00 | 2.13 | 0.99 | | | 2.03 | 1.00 | 2.21 | 0.99 | | |
| 50 | 3.78 [b] | 1.00 | 3.72 | 0.97 | 1.68** [c] | 0.94 | 3.46 | 1.00 | 3.17 | 0.98 | 1.66** [c] | 0.89 |

[a] Unit of *k'* is $cm^2\ J^{-1}\ (\frac{C_0/C}{Fluence})$; ** shows 95% significant differences compared with HPLC or YES analyses at the same treatments; *shows 90% statistically significant difference.
[b] Regression for UV fluence between 0-3000 mJ $cm^{-2}$ based on detection limits
[c] Regression for UV fluence between 0-1000 mJ $cm^{-2}$ based on detection limits
[d] LP-UV data are from Chen et al., 2006.

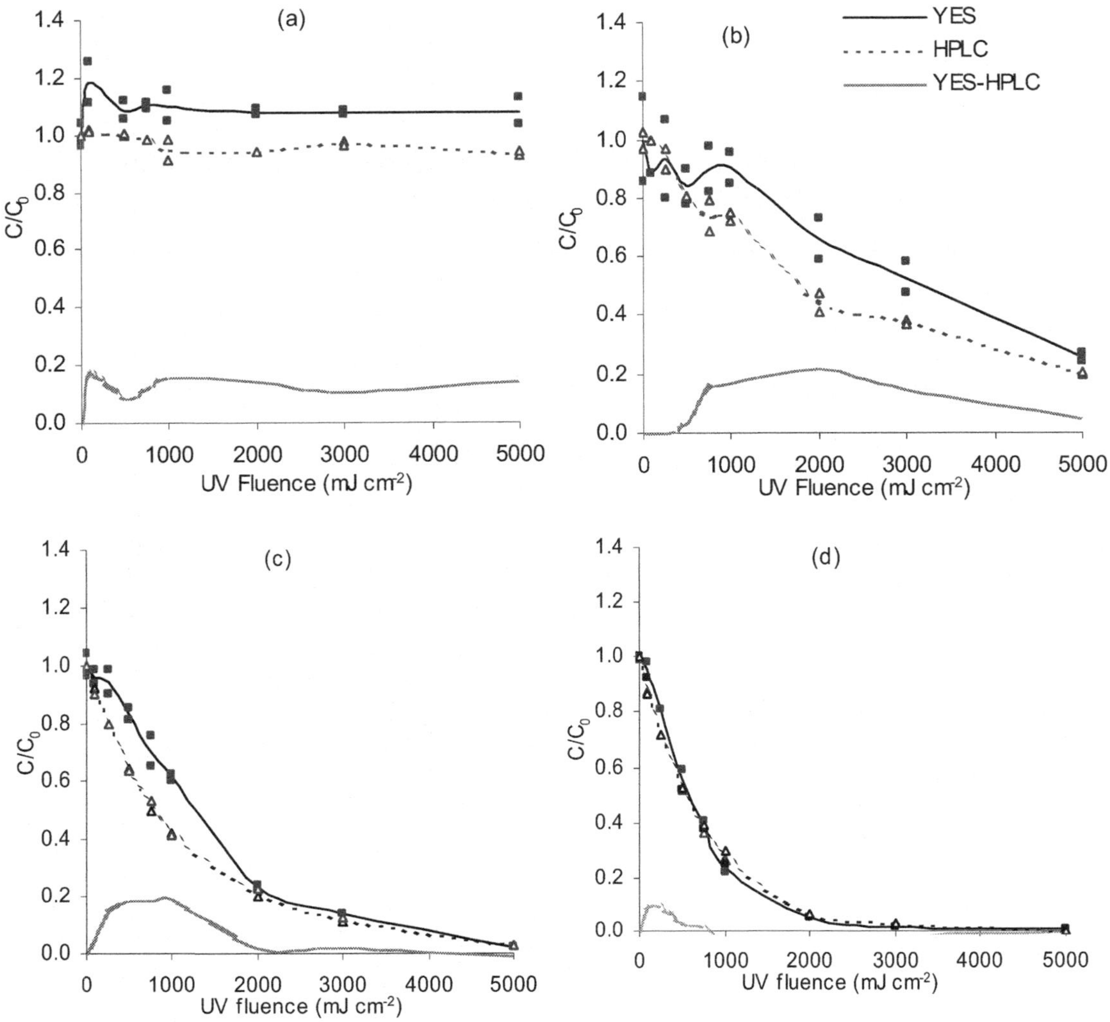


**Figure 4.1. Analyses of BPA degradation (HPLC analysis) and estrogenic activity (YES assays) with direct photolysis with LP lamps (a) and LP-UV/AOP in combination of 10 ppm $H_2O_2$ (b), 25 ppm $H_2O_2$ (c), and 50 ppm $H_2O_2$ (d). Black lines represent the average of estrogenic activity (full square). Dashed lines represent the average of parent BPA (empty triangles). Gray dashed lines show differences between two averages. Source: Reprinted from *Chemosphere*, 65, Chen, P.J., K.G. Linden, D.E. Hinton, S. Kashiwada, E.J. Rosenfeldt, and S.W. Kullman. Biological assessments of Bisphenol A degradation in water following direct photolysis and UV advanced oxidation. 1094–1102, Copyright 2006, with permission from Elsevier.**

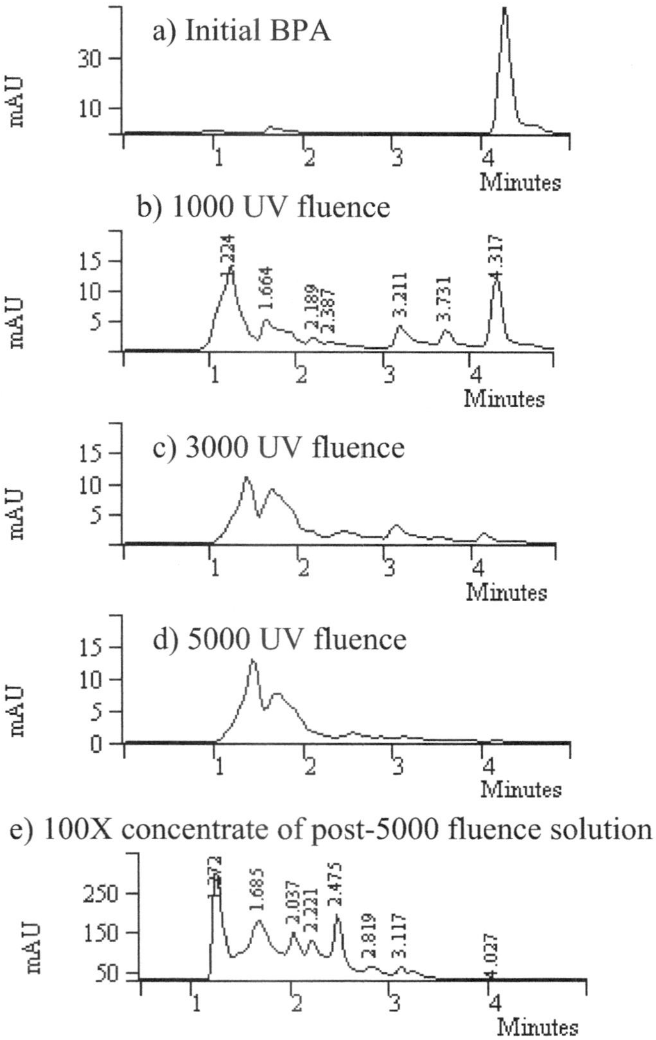


**Figure 4.2. HPLC chromatogram of BPA degradation products following treatment of 60 μM BPA with LP-UV/AOP + 50 ppm $H_2O_2$. Reprinted from *Chemosphere*, 65, Chen, P.J., K.G. Linden, D.E. Hinton, S. Kashiwada, E.J. Rosenfeldt, and S.W. Kullman. Biological assessments of Bisphenol A degradation in water following direct photolysis and UV advanced oxidation. 1094–1102, Copyright 2006, with permission from Elsevier.**

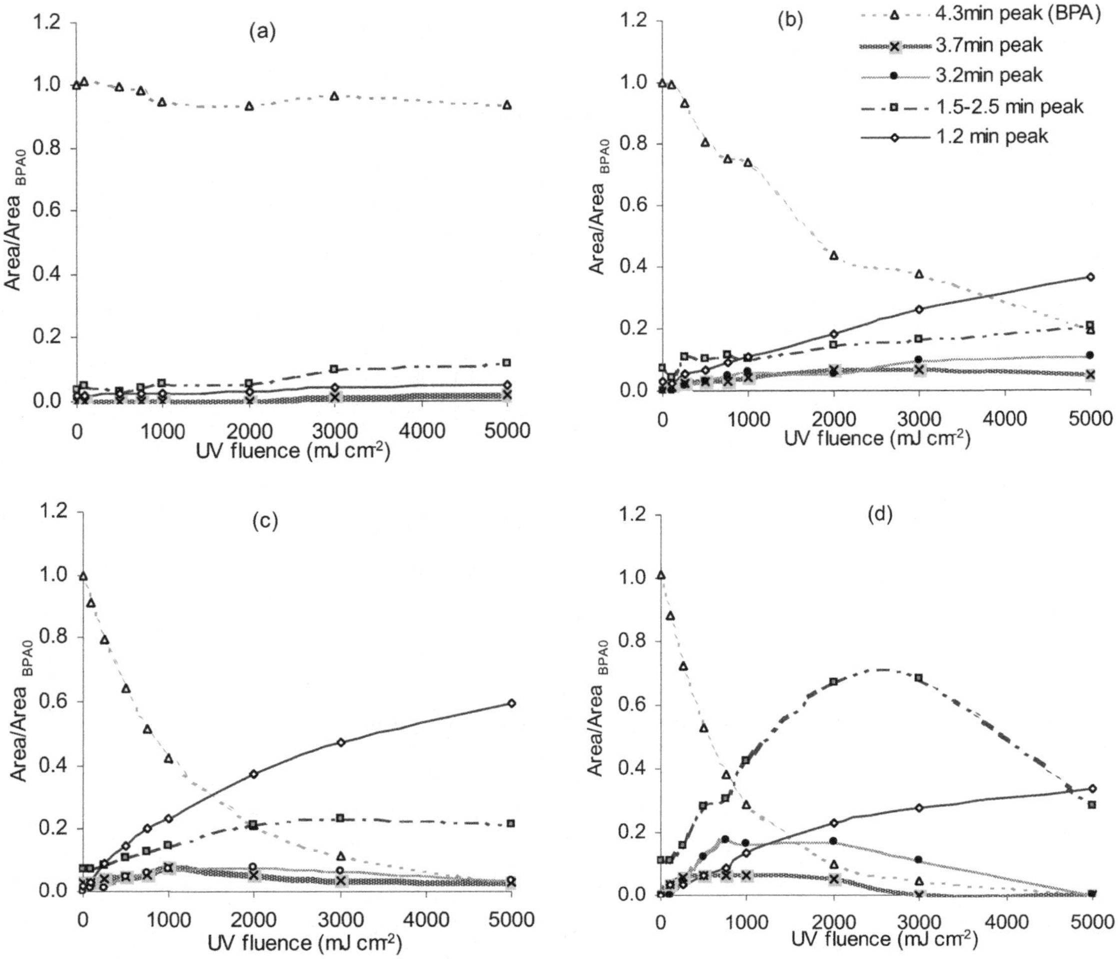


**Figure 4.3. HPLC analysis of BPA transformations following direct LP-UV (a) or UV/AOP with addition of 10 ppm $H_2O_2$ (b), 25 ppm $H_2O_2$ (c), and 50 ppm H2O2 (d). Retention times for BPA and individual degradation products are listed in the legend of (b). The area of each peak was normalized by the peak area of initial BPA as Area/Area$_{BPA0}$. Data represent the mean of duplicates. Matched analysis of YES data is given in Figure 4.1. Reprinted from *Chemosphere*, 65, Chen, P.J., K.G. Linden, D.E. Hinton, S. Kashiwada, E.J. Rosenfeldt, and S.W. Kullman. Biological assessments of Bisphenol A degradation in water following direct photolysis and UV advanced oxidation. 1094–1102, Copyright 2006, with permission from Elsevier.**

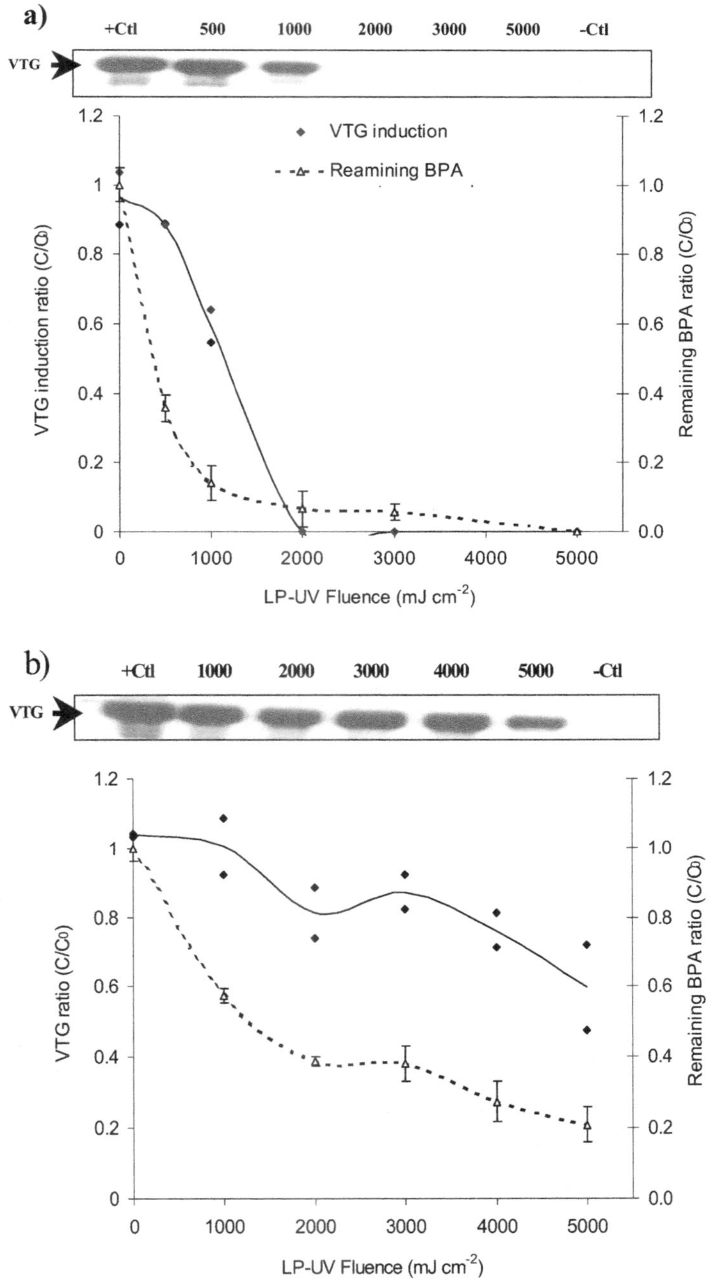


**Figure 4.4.** ***In vivo*** **VTG induction following a one week aqueous exposure of post LP-UV/$H_2O_2$ BPA solutions. (a) LP-UV/50-ppm $H_2O_2$ treatment; (b) LP-UV/10-ppm $H_2O_2$ treatment. Number above gel images indicates UV fluence; + Ctl is a positive control, and - Ctl is a blank control. 60 μM of initial BPA solution resulted in 100% fish mortality by 24 hours, so $C_0$ is substituted with + Ctl. Bold lines represent the average of VTG induction ratio ($C/C_0$) of three biological replications (Full squares). Full triangle represents the average of 7d BPA concentration in the dosing solution with standard deviation in error bars. Reprinted from *Chemosphere*, 65, Chen, P.J., K.G. Linden, D.E. Hinton, S. Kashiwada, E.J. Rosenfeldt, and S.W. Kullman. Biological assessments of Bisphenol A degradation in water following direct photolysis and UV advanced oxidation. 1094–1102, Copyright 2006, with permission from Elsevier.**

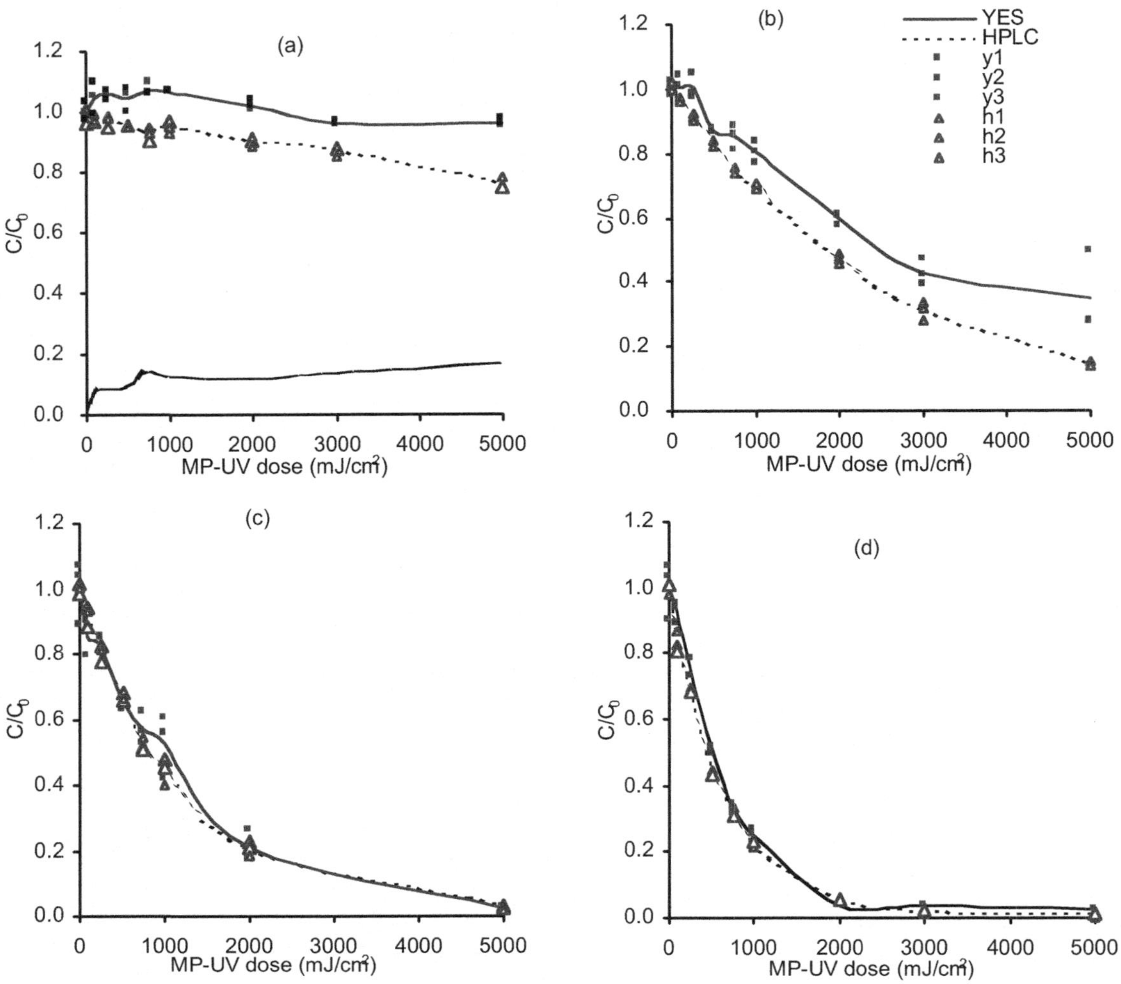


**Figure 4.5. Analyses of BPA degradation and estrogenic activity with direct photolysis of MP-UV lamps (a), and MP-UV/AOP in combination with 10 ppm $H_2O_2$ (b), 25 ppm $H_2O_2$ (c), and 50 ppm $H_2O_2$ (d) with the use of MP-UV lamps. Black lines represent the average estrogenic activity of three replicates (full square). Dashed lines represent the average parent BPA concentration of three replicates (empty triangle).**

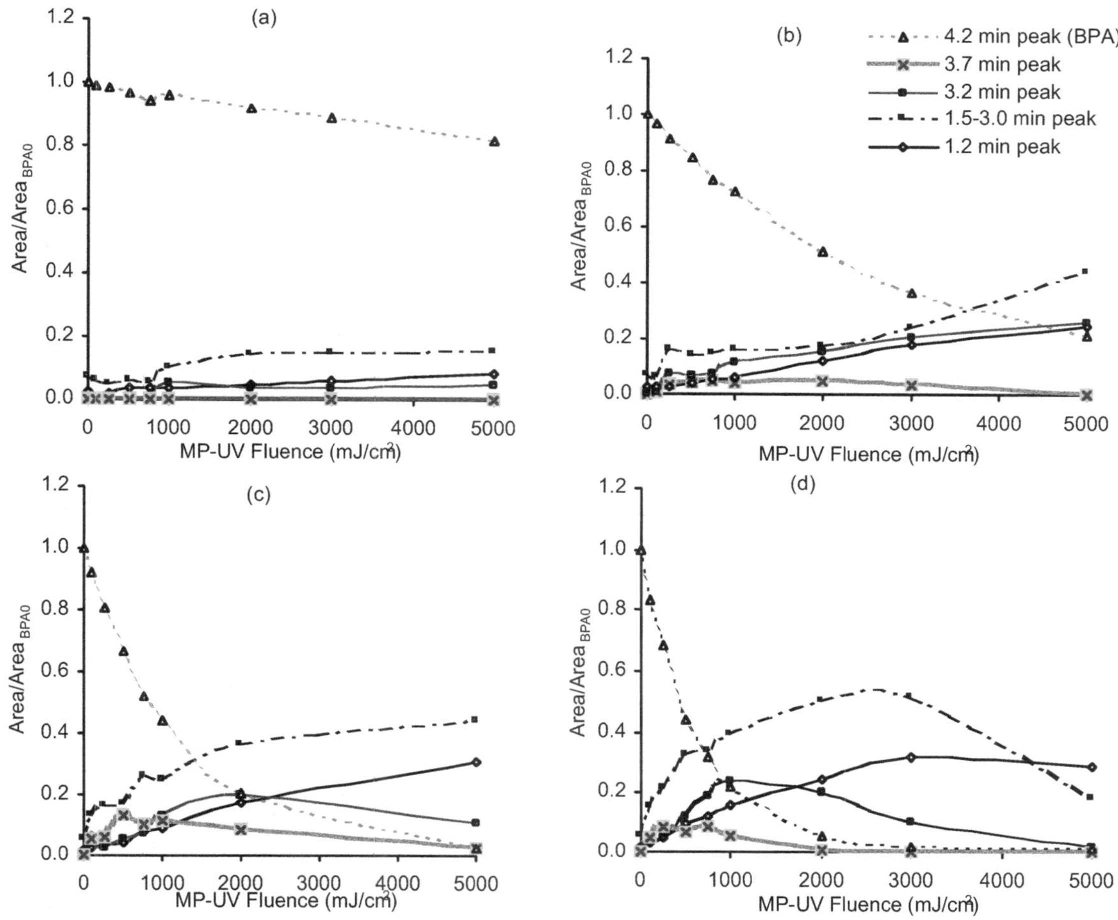


**Figure 4.6. HPLC analysis of BPA transformations following direct MP-UV (a) or UV/AOP with 10 ppm $H_2O_2$ (b), 25 ppm $H_2O_2$ (c), and 50 ppm $H_2O_2$ (d). Retention times for BPA and individual degradation products are listed in the legend of (b). Area of each peak was normalized by the peak area of initial BPA as Area/Area$_{BPA0}$. Data represent the mean of three replicates. Matched analysis of YES data is given in Figure 4.2.**

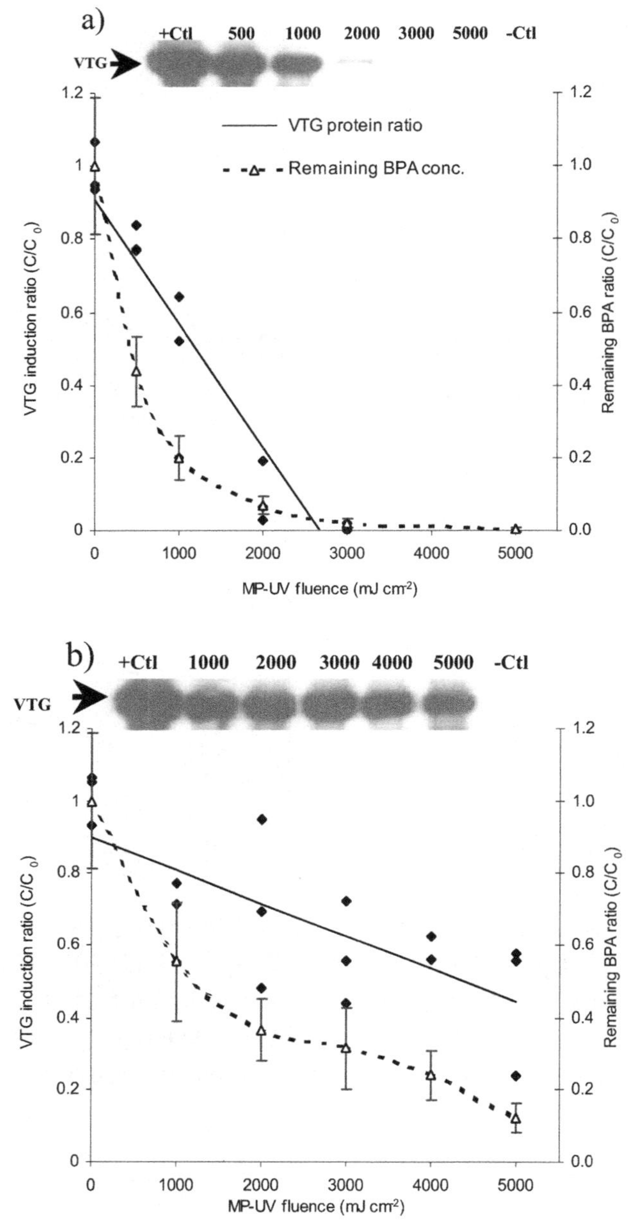


**Figure 4.7** ***In vivo*** **VTG induction following one week of aqueous exposure of post MP-UV/$H_2O_2$ BPA solutions. (a) MP-UV/50-ppm $H_2O_2$ treatment; (b) MP-UV/10-ppm $H_2O_2$ treatment. Number above gel images indicates UV fluence; +Ctl is a positive control, and -Ctl is a blank control. 60 μM of initial BPA solution caused complete fish death in 24 hours, so $C_0$ is substituted with +Ctl. Solid lines represent the linear regression of VTG induction ratio ($C/C_0$) of three biological replications (Full diamond). Dashed lines represent the average of 7-day BPA concentration in the dosing solution with standard deviation in error bars.**

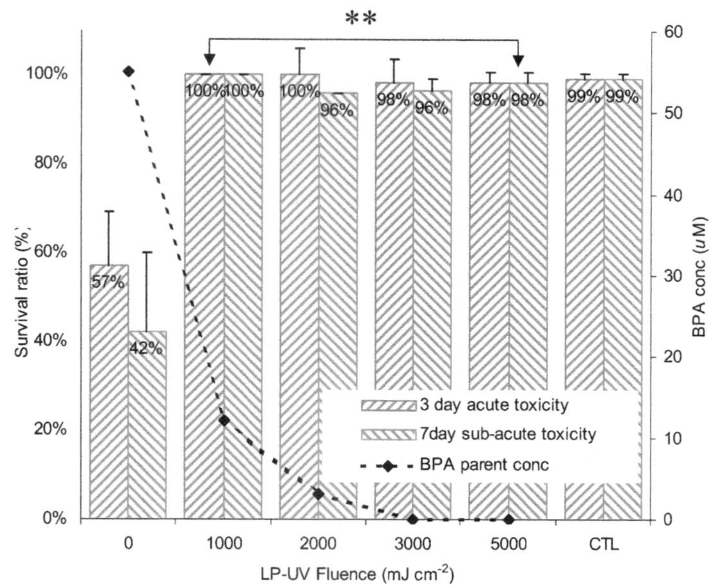


(a)

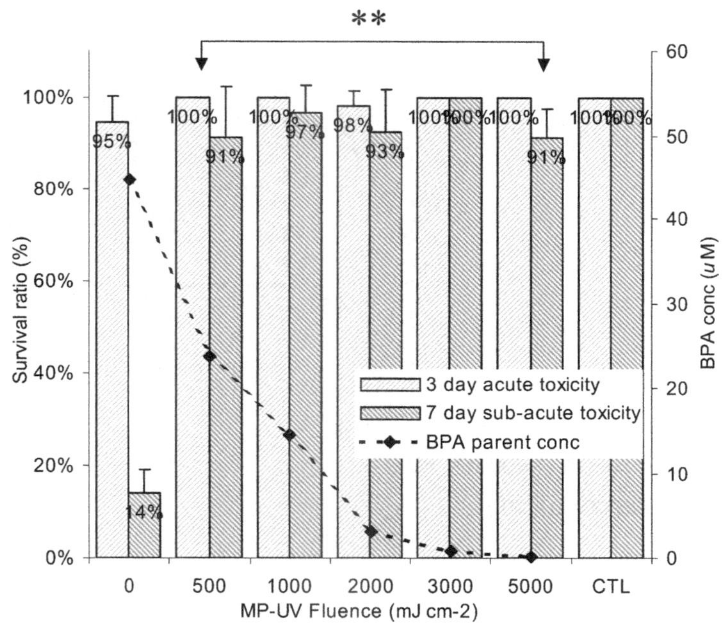


(b)

**Figure 4.8 Larval survival ratio following the exposure of treated water under (a) LP-UV; and (b) MP- UV irradiation + 50ppm- $H_2O_2$ treatments for 60 μM BPA solution. ** indicates 95% of significant differences compared to initial sample prior to UV irradiation for both acute and sub-acute toxicity.**

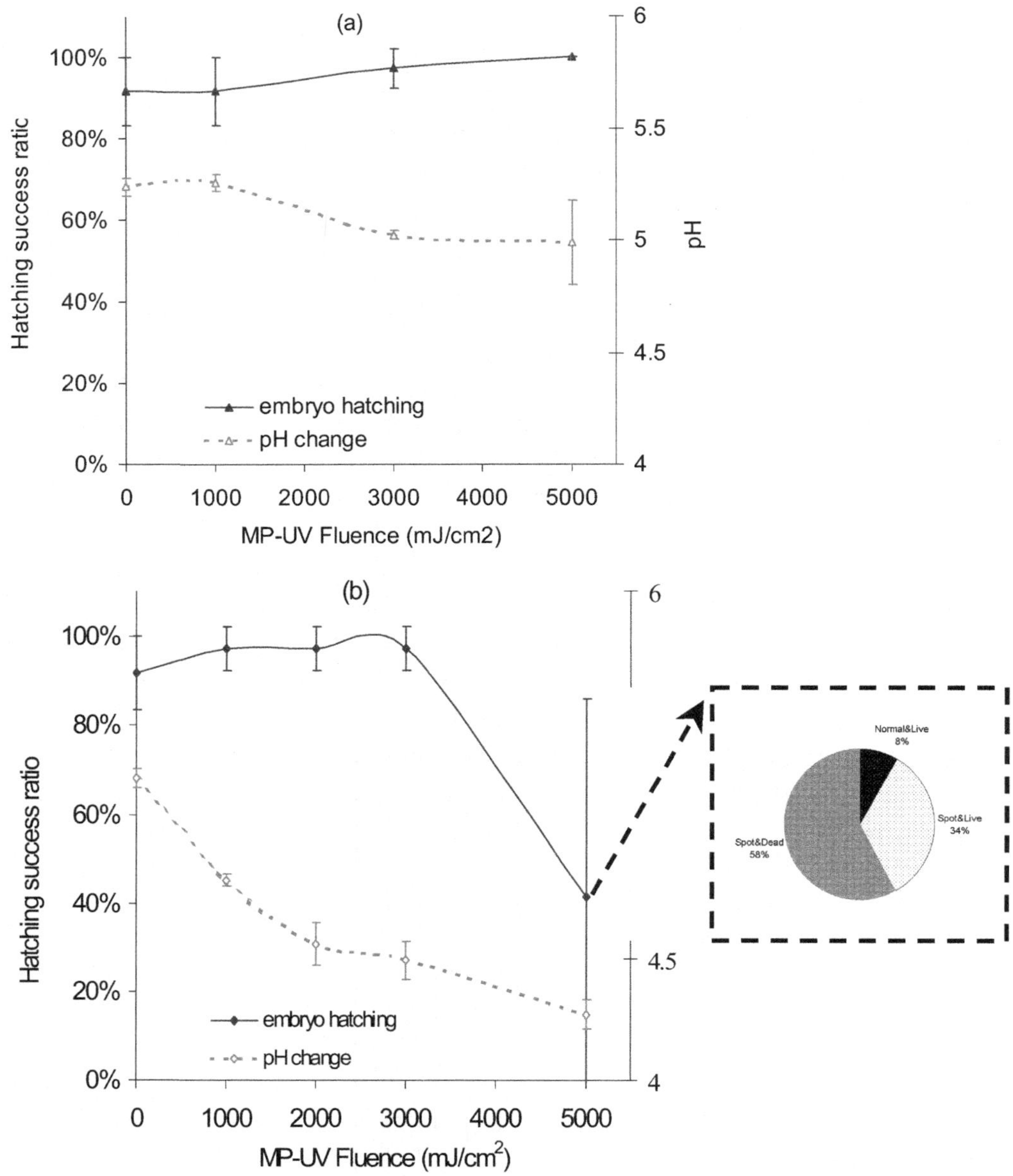


**Figure 4.9 Hatching success ratio of embryo under two UV/AOP treatments with addition of (a) 10- and (b) 50- ppm $H_2O_2$ for 60 μM BPA solutions. Data are presented as mean with SD for three replicates of UV/AOP experiments. 12 embryos each replicate for a total of 36 embryos each treatment were used.**

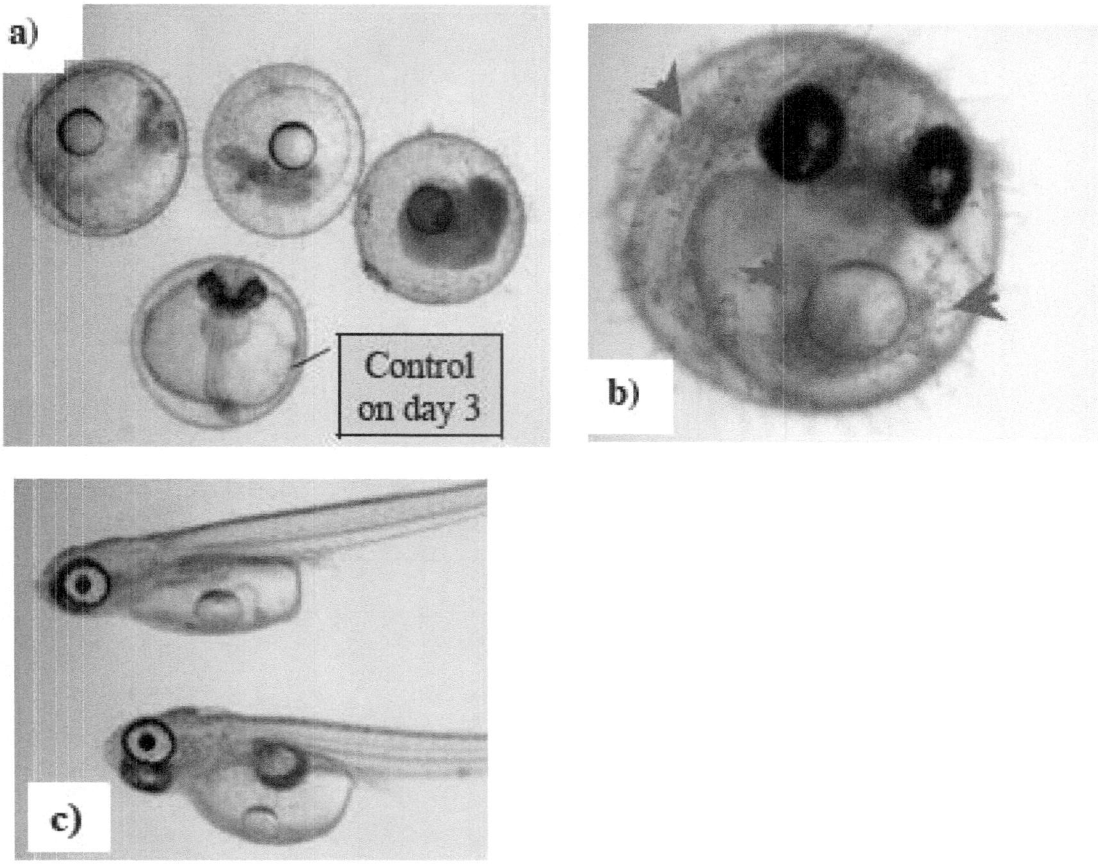


**Figure 4.10 Gross morphologies of embryos following exposure of BPA solution after MP-UV/50-ppm $H_2O_2$ treatment at 5000 mJ/cm$^2$ (a) embryos at day 3 in treated solution (upper row) and the control (bottom one ) (b) embryos at day 7 with granulated spots inside the chorion (arrowhead pointed) (c) hatchling at day 14 with inflated swim bladder (bottom) or without (up) after exposure.**

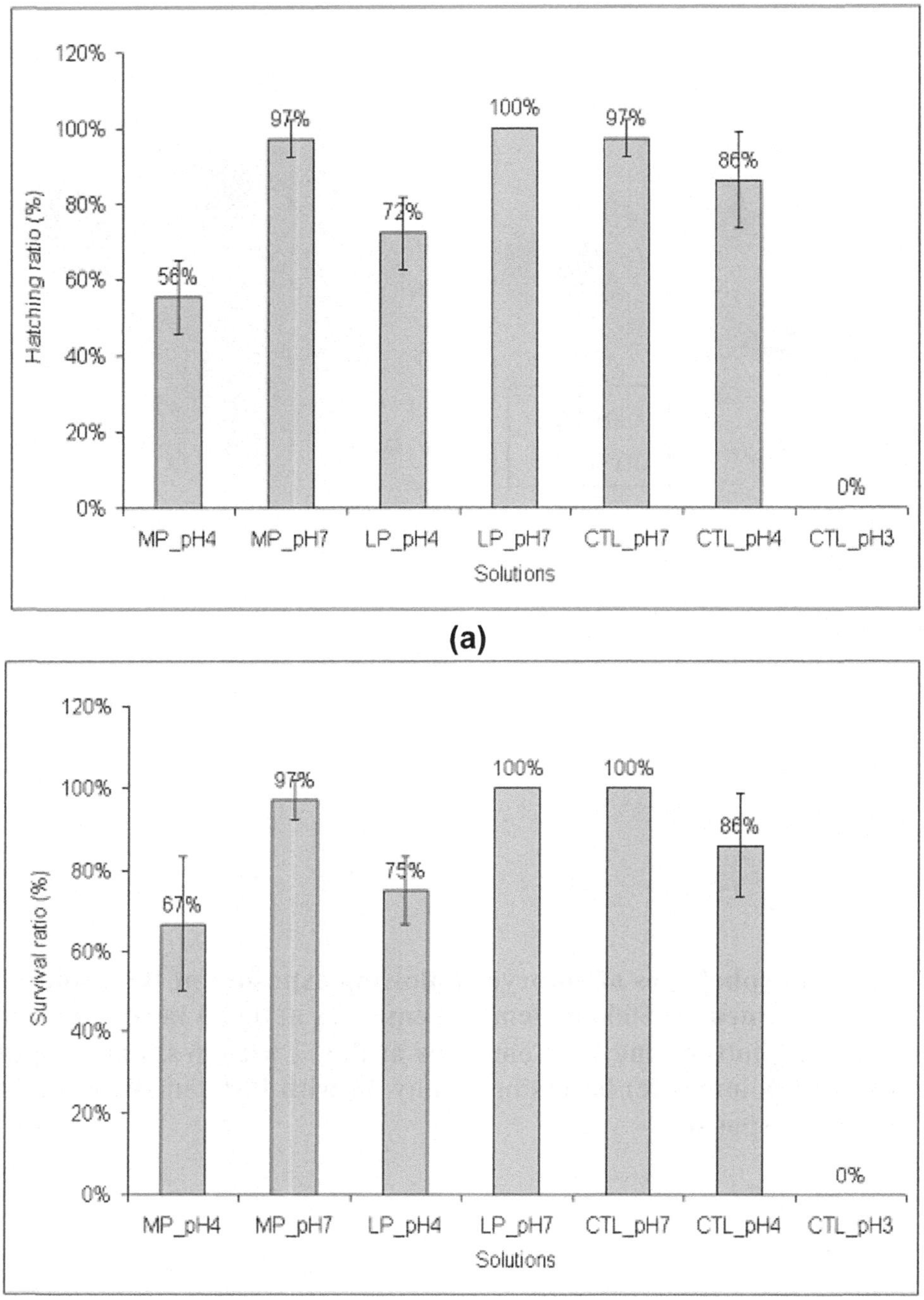


(b)

**Figure 4.11 (a) Hatching success ratio, and (b) sub-acute toxicity of embryos at different pH values of BPA-treated solution or blank controls (ERM). 60 μM BPA was treated with LP- and MP-UV+50ppm $H_2O_2$ at 5000 mJ/cm2 without (pH=4) or with adjusted pH (to pH=7) for the treated solutions. Three pH of ERM control (CTL_pH7, CTL_pH4 and CTL_pH3) were used. Triplicates each treatment for a total of 36 embryos were used. Error bars: SD.**

# CHAPTER 5
# BIOANALYTICAL ANALYSIS OF UV TREATMENT OF EE2 AND E2

## INTRODUCTION

The physical removal or chemical transformation destruction of steroidal EDCs by various processes, including membranes (Kimura et al., 2004; Yoon et al., 2004, Nghiem et al., 2004), adsorption (Yoon et al., 2005), oxidation by chlorine dioxide (Huber et al., 2005) and ozone (Huber et al., 2003; Huber et al., 2004; Irmak, 2005), has been well documented. Reviews by Snyder et al. (2003) and Westerhoff et al. (2005), qualitatively and quantitatively detail how effective various water treatment unit operations are at removing or transforming EDCs in water. Advanced oxidation processes, including $UV/O_3$ (Westerhoff et al. 2005; Huber et al., 2004), $UV/H_2O_2$ (Rosenfeldt and Linden, 2004), and $UV/TiO_2$ photocatalysis (Ohko et al., 2001; Ohko et al., 2002; Nakashima et al., 2002), have been determined successful at oxidizing natural and synthetic estrogens. In Chapter 3, we demonstrated the ability of UV and $UV/H_2O_2$ for degrading 17-β-estradiol (E2) and 17-α-ethinyl estradiol (EE2), but noted that of ultimate concern in any treatment process is an assessment of the treatment success in elimination of the toxic action of the parent pollutant. Because complete mineralization of E2 and EE2 to carbon dioxide and water is impractical due to extended UV treatment times and elevated peroxide levels, remaining estrogenic activity associated with oxidation products formed during $UV/H_2O_2$ are of concern.

Several studies have reported concerning results regarding oxidation of several EDCs and an increase in estrogenic activity. Hu et al. (2002a,b) found that chlorinating bisphenol A (BPA) and 4-nonylphenol (4-NP) led to increased estrogenic activity as measured by a yeast two-hybrid assay system, and Lenz et al. (2004) found similar results chlorinating BPA and 4-NP measuring with the YES. Fortunately, when this study examined stronger oxidants, chlorine dioxide and ozone, the removal of estrogenic activity corresponded to BPA and 4-NP oxidation. In Chapter 4 reporting work examining oxidation and the corresponding removal of estrogenic activity associated with BPA via $UV/H_2O_2$ AOP (also see Chen et al, 2006), a discrepancy between removal of estrogenic activity and parent compound was reported, suggesting the present of some oxidation byproducts that may retain some estrogenic activity. Here the estrogenic activity associated with products of $UV/H_2O_2$ advanced oxidation of E2 and EE2 as measured by the *in vitro* Yeast Estrogen Screen (YES) is reported.

First the correlation between removal of estrogenic activity and oxidation of E2 and EE2 was examined, then the YES was utilized as an analytical tool to follow the destruction of E2 and EE2 at environmentally relevant concentrations. These compounds have been found at μg/L concentrations in surface waters throughout the United States and the world (Koplin et al., 2002; Ying et al., 2002; Jurgens et al., 2002), and utilizing the YES as an analytical tool allows us to examine the kinetics of E2 and EE2 removal at these environmental levels. The work was performed in lab water, as well as several natural waters, to examine the effects of water quality on removal of estrogenic activity. The $UV/H_2O_2$ AOP is modeled by combining the steady state OH radical model (Sharpless and Linden, 2003) to determine EDC concentration, and variable slope, sigmoidal functions to relate this concentration to a YES response.

## MATERIALS AND METHODS

### Chemicals and Waters

E2 and EE2 were purchased as solids from Steraloids, Inc., Newport, RI. The chemicals used for the mobile phase of HPLC detection included HPLC grade Acetonitrile (Burdick and Jackson, Muskegon, MI), and water purchased from Malinkrodt Chemical Company (St. Louis, MO). Lab water was deionized and purified water taken from a laboratory water purifying system (Hydro Services and Supplies, Inc., Research Triangle Park, NC). Two natural water samples were taken from water treatment plant inlets, before any treatment. Samples were then filtered through a 0.45μm filter to remove particles, and water quality parameters including DOC, alkalinity, pH, and UV absorbance were measured to characterize the waters.

### Chemical Analytical Equipment and Methods

A Varian Pro Star High Performance Liquid Chromatography (HPLC) workstation (Walnut Creek, CA), with a model 330 Polychromatic Diode Array detector was used to detect the concentration of EDCs in water. A C-18 reverse phase column served as the stationary phase, with 1 mL $min^{-1}$ of eluent consisting of an acetonitrile (ACN) to water ratio of 60%:40% for EE2, and 50%:50% for E2. These HPLC methods were adapted from the literature for EE2 (Huber et al., 2003) and E2 (Ohko et al., 2002). Water quality parameters were measured using standard methods. Total organic carbon (TOC) was measured in accordance with Standard Method 5310 A (combustion and detection method) (Tekmar Dohrmann Apollo 9000 Total Carbon analyzer). Alkalinity was measured according to Standard Methods 2320 B. Residual $H_2O_2$ was measured using the $I_3^-$ method (Klassen et al., 1994). pH was measured using an electronic pH meter (Cole Parmer pH 100 series), calibrated daily using pH 4, 7, and 10 buffers. UV absorbance (200-300 nm) of the spiked test water was measured in a spectrophotometer (Cary 100 bio-spectrophotometer, Varian, Houston, TX).

### Yeast Estrogen Screen (YES)

The YES assay was performed according to the method by Routledge and Sumpter (1996) with modifications as described in Chen et al., (2006). Briefly, 100 μl dilutions of E2 and EE2 standards (0-100 μM) and post treated solutions were added to 96 deep-well plates (1 mL capacity/well, VWR, West Chester, PA) containing 300 μl of diluted yeast solution an optical density (O.D.) of 0.08 at 630 nm. Plates were covered with Kimwipes papers (Kimtech science, Roswell, GA) and incubated at 30°C for 3 days with gentle shaking. 17β-estradiol (0.0001- 10 nM) served as a positive control. After 3 days of incubation, 400 μl of assay buffer (containing ortho-nitrophenyl-beta-D-galactopyranoside (ONPG) substrate and lysis buffer) was then added to each well. Plates were sealed with an adhesive seals membrane (USA Scientific Inc., Ocala, FL) and incubated at 30 $^0$C for 20 minutes. 200 μl of 1 M sodium carbonate was used to stop β-galactosidase reactions in the media. Plates were centrifuged at 3,000 rpm for 10 minutes. To determine O.D. at 405 nm and 630 nm, 100 μl of supernatant was taken from each well and then transferred to a new 96 well microtiter plate. All values were presented as OD405 minus OD630.

### UV Reactors and Fluence Calculation

Irradiation by a 1 kW medium pressure UV lamp (Hanovia Co., Union, New Jersey) was performed using a bench-scale collimated beam UV reactor provided by Calgon Carbon Corporation (CCC, Pittsburgh, Pennsylvania) modified at Duke University. Irradiation by a low pressure UV lamp was performed using another bench-scale collimated beam UV reactor fitted with four 15-W low-pressure UV lamps. A typical 'collimated beam' UV system consists of a light source, a 'collimating' device, a support stand, and an exposure control system comprising a manual (LP UV) or pneumatic (MP UV) shutter, and has been described in detail in Chapter 2. UV fluence (mJ $cm^{-2}$, where J = W s) was calculated as the average irradiance multiplied by the exposure time. The average UV fluence rate (mW $cm^{-2}$) in the completely mixed sample was determined from the incident irradiance, UV absorbance and sample depth using an integrated form of the Beer-Lambert law and calculated using an MS Excel spreadsheet (Bolton and Linden, 2003). Incident UV irradiance (mW $cm^{-2}$) was measured using A UV radiometer and detector (International Light Inc., Model 1700/SED 240/W) calibrated traceable to standards at the National Institute of Standards and Technology (NIST) at 2 nm intervals in the range of 200 to 400 nm. All UV fluence values reported for the MP UV source were calculated as a function of the entire lamp output between 200 and 300 nm.

## EXPERIMENTAL DESIGN

The experiments were performed in two stages. First, experiments were performed to examine the link between UV/$H_2O_2$ oxidation of the parent EDC and corresponding reduction in estrogenic activity, as measured by the YES. These experiments were performed at concentrations of approximately 5μM initial EDC, and oxidation was performed with either LP UV + 5 mg $L^{-1}$ $H_2O_2$, MP UV + 5 mg $L^{-1}$ $H_2O_2$, or MP UV + 10 mg $L^{-1}$ $H_2O_2$. Seven UV fluences of 0, 200, 500, 750, 1000, and 1500 mJ $cm^{-2}$ were applied, and samples were taken for HPLC and YES analysis. Samples destined for YES analysis were diluted with the same dilution factor, calculated such that the initial EDC concentration would be diluted to $1x10^{-8}$ M.

To examine removal of estrogenic activity associated with environmentally relevant concentrations of E2 and EE2 in lab water as well as several natural waters, initial EDC concentrations were maintained at 10nM. Treatment conditions again included LP UV + 5 mg $L^{-1}$ $H_2O_2$, MP UV + 5 mg $L^{-1}$ $H_2O_2$, or MP UV + 10 mg $L^{-1}$ $H_2O_2$. Twelve UV UV fluences were applied, including 0, 50, 100, 150, 200, 250, 300, 350, 400, 500, 600, and 700 mJ $cm^{-2}$, and samples were collected for YES analysis. YES analysis was performed as referenced in a 96 well plate, and generic plate maps for each set of experiments are displayed in Tables 5.1 and 5.2.

In the oxidation product estrogenic activity experiments (Table5. 1), 200 μL of the initial diluted sample for each UV fluence was placed in column 1. 100 μL of this was then removed and mixed with 100 μL of DI water in column 2. The process was repeated for all 12 columns, diluting 1:2 with each column. In the environmentally relevant concentration experiments (Table 5.2), 100 μL of sample collected at each UV fluence was placed in the corresponding cell.

## RESULTS AND DISCUSSION

### Direct UV Photolysis Products: Residual Estrogenic Activity

YES data from "table 1 type" experiments were analyzed using Prism software (Graphpad, San Diego) based on fitting the Hill equation. For the analysis, raw absorbance values were converted to relative absorbance, and a log transform of the dilution factors was performed. Upon fitting with the Hill equation, an $EC_{50}$ dilution factor was obtained. Plots of the natural log transform of the ratio of $(EC_{50}/EC_{50}^{o})$ where $EC_{50}^{o}$ corresponds to 0 mJ $cm^{-2}$ as a function of UV fluence for each treatment process yielded pseudo-first order rate constants of estrogenic activity removal which are compared to the pseudo-first order rate constant describing E2 or EE2 photolysis in Figure 5.1.

There is no statistically significant difference between photolysis of E2 and EE2 and the corresponding reduction in estrogenic activity, within the 95% confidence interval. Because of this, it can be assumed that the observed degradation in parent compound corresponds directly with a reduction in estrogenic activity. The differences observed in photolysis rates of E2 and EE2 between the MP and LP UV systems correspond with previously reported trends in the quantum yield values. In Chapter 3, we report quantum yield values for MP UV of approximately double those of LP UV for E2 and EE2, and as observed in Figure 5.1, these trends can also be extended to the reduction in estrogenic activity of these EDCs by direct UV photolysis.

### UV/$H_2O_2$ Oxidation Products: Residual Estrogenic Activity

UV/$H_2O_2$ AOP experiments were analyzed in the same manner as the direct UV photolysis experiments, and the data is shown in Figure 5.2.

Within the 95% confidence interval, there is no statistically significant difference between the rates of oxidation and rate of removal of estrogenic activity for E2 or EE2 treated with the three AOP conditions. This result implies two things. First, none of the oxidation byproducts produced by UV/$H_2O_2$ AOP are comparable with the parent compound in terms of estrogenic activity. Secondly, it implies that for E2 and EE2, YES can be utilized as an analytical tool for measuring oxidation behavior at low, environmentally relevant concentrations.

### Comparison of UV photolysis and UV/$H_2O_2$ AOP

For "Table 5.2" type experiments, relative YES absorbance data was plotted as a function of applied UV fluence to observe the degradation of estrogenic activity. Figure 5.3 displays the removal of estrogenic activity of EE2 spiked at 3 $\mu g\ L^{-1}$ in DI water using direct LP or MP photolysis, compared to UV + 5 mg $L^{-1}$ $H_2O_2$ as a function of UV fluence.

Inspection of Figure 5.3 implies that LP UV is not capable of reducing estrogenic activity associated with EE2 with an applied fluence of 14,000 mJ $cm^{-2}$. However, MP UV is capable of degrading estrogenic activity associated with EE2 by 80% at an applied UV fluence of 5,000 mJ $cm^{-2}$. For comparison, in Coleman et al (2004) it was shown that UVA (from a 125 Watt Philips High-Pressure Mercury lamp) could degrade E2 and EE2, although slowly compared to UVA + $TiO_2$. The authors showed that addition of titanium dioxide to the system would increase the rate of destruction by 9.6 times for E2 and 2.5 times for EE2. In contrast, the addition of 5 mg $L^{-1}$ $H_2O_2$ to the MP UV system increases the rate of E2 destruction by 133 times and the rate of EE2 destruction by 27.5 times.

### Water Quality Effects on EE2 Oxidation by UV/$H_2O_2$ AOP

Figure 5.4 displays removal of estrogenic activity associated with 3μg $L^{-1}$ EE2 in DI water and two natural waters with the MP UV + 5mg $L^{-1}$ $H_2O_2$ AOP. Water 1 contains 3.1 mg $L^{-1}$ Dissolved Organic Carbon (DOC), 2.63 mM alkalinity, 2.97 mg $L^{-1}$ Nitrate, pH 8.0 and $UV_{254}$ of 0.04 $cm^{-1}$. Water 2 contains 1.39 mg $L^{-1}$ DOC, 0.67mM alkalinity, 1.30 mM Nitrate, pH 8.0 and $UV_{254}$ of 0.11 $cm^{-1}$. UV/$H_2O_2$ AOP oxidation of EDCs is greatly affected by scavenging. Assuming the following rate constants for the reaction between OH radical and the various water quality parameters: DOC ($2.5x10^4$ L $mg^{-1}$ $s^{-1}$) (30), $HCO_3^-$ ($8.5x10^6$ $M^{-1}$ $s^{-1}$), $CO_3^{-2}$ ($3.9x10^8$ $M^{-1}$ $s^{-1}$), and $H_2O_2$ ($2.7x10^7$ $M^{-1}$ $s^{-1}$) (Buxton et al., 1988), the theoretical overall scavenging factors in each water matrix was calculated as $4.0x10^3$ $s^{-1}$ for DI water (all from $H_2O_2$ scavenging), $3.6x10^5$ $s^{-1}$ for water 1 and $4.0x10^4$ $s^{-1}$ for water 2. This theoretical scavenging does not accurately explain the results in Figure 4, as it implies the kinetics should be faster in water 2 than water 1. However, other factors, such as nitrate, which is known to generate OH radicals (Sharpless and Linden, 2001), and NOM sensitized OH radical formation are not accounted for in this scavenging term. Also, the theoretical DOC scavenging term is a lumped parameter utilized to describe the OH radical scavenging ability of NOM, and does not account for variability in the structure and reactivity of NOM in different waters. Finally, the background absorbance in water 1 is less than in water 2, implying more of the total absorbed UV energy is utilized in producing OH radicals in water 2 than water 1.

### Modeling of Estrogenic Activity Removal

The steady state OH radical model, introduced by Glaze et al. (1995), has been successfully utilized to model the degradation of several contaminants, including NDMA (Sharpless and Linden, 2003), MIB and geosmin (Rosenfeldt et al., 2005), and EDCs (Rosenfeldt and Linden, 2004; 2005). However, prediction of the removal of estrogenic activity associated with these EDCs has not been explored. The approached utilized in this work involved first calculating the EE2 concentration via the steady state OH radical method, and converting this concentration to an estrogenic response with a model fit to EE2 calibration curves.

Equations 1-5 from Rosenfeldt et al. (2005) were used to determine [EE2] at each applied UV fluence, accounting for the scavengers (and corresponding OH radical rate constants) organic carbon ($2.5x10^4$ L $mg^{-1}$ $s^{-1}$) (Larson and Zepp 1988), $HCO_3^-$ ($8.5x10^6$ $M^{-1}$ $s^{-1}$), $CO_3^{-2}$ ($3.9x10^8$ $M^{-1}$ $s^{-1}$), and $H_2O_2$ ($2.7x10^7$ $M^{-1}$ $s^{-1}$) (Buxton et al., 1988). Once [EE2] was determined, the relative YES response was calculated using a variable-slope sigmoidal function (equation A) fit from Graphpad Prism software.

$$\frac{A}{A_o} = B + \left\{ \frac{T - B}{1 + 10^{(\log(EC_{50}) - \log[EE2])H}} \right\} \qquad \text{(A)}$$

where $\frac{A}{A_o}$ is the relative YES response, B is the minimum relative YES response, T is the maximum relative YES response, $EC_{50}$ is the [EE2] associated with 50% relative YES response, and H is the hillslope describing the "steepness" of the change response with respect to changing [EE2]. All of these parameters are found for a specific calibration curve in the water to be modeled, and are displayed in Table 5.3.

Results comparing the model predicted oxidation and actual data for LP + 5, MP + 5, and MP + 10 mg $L^{-1}$ $H_2O_2$ experiments are shown in Figure 5.5.

The model slightly underpredicts the actual oxidation, but the general shape of the curves are similar, and both the model and data predict 1 log removal of estrogenic activity at UV fluence of less and 200 mJ $cm^{-2}$. To compare the predicted and actual oxidation rates for all waters, rates of removal of estrogenic activity were compared in Figure 5.6.

While the rates are comparable in DI water, the model under-predicted oxidation rates in the natural waters. There are several possible explanations for this discrepancy, including indirect NOM photosensitized degradation of the EDCs, as well as photolysis of nitrate to OH radical. Also, the assumption of TOC as an acceptable NOM surrogate and the lumped OH radical rate constant for TOC, the theoretical NOM scavenging may be over-predicting the actual NOM contribution to the overall scavenging in the natural waters. Additionally, there may be unknown interactions between NOM and the estrogenic activity response as measured by the YES that lead to an enhanced removal of the observed estrogenic activity. These interactions are not well understood, and recognized as areas of future research.

## Table 5.1

**Plate map for the oxidation product estrogenic activity studies. The values in the cells of row A are the calibration concentrations, while in rows B-H, the values in the cells are the dilution factor tested.**

| | 1 | 2 | 3 | 4 | 5 | 6 | 7 | 8 | 9 | 10 | 11 | 12 |
|---|---|---|---|---|---|---|---|---|---|---|---|---|
| A (Calibration) | 1.E-08 | 5.E-09 | 1.E-09 | 8.E-10 | 5.E-10 | 2.E-10 | 1.E-10 | 8.E-11 | 5.E-11 | 1.E-11 | DI | DI |
| B (0 mJ cm-2) | 500 | 1000 | 2000 | 4000 | 8000 | 16000 | 32000 | 64000 | 128000 | 256000 | 512000 | 1024000 |
| C (100 mJ cm-2) | " | " | " | " | " | " | " | " | " | " | " | " |
| D (200 mJ cm-2) | " | " | " | " | " | " | " | " | " | " | " | " |
| E ( 500 mJ cm-2) | " | " | " | " | " | " | " | " | " | " | " | " |
| F (750 mJ cm-2) | " | " | " | " | " | " | " | " | " | " | " | " |
| G (1000 mJ cm-2) | " | " | " | " | " | " | " | " | " | " | " | " |
| H (1500 mJ cm-2) | " | " | " | " | " | " | " | " | " | " | " | " |

## Table 5.2

**Plate map for the environmentally relevant UV/$H_2O_2$ experiments. Values in rows A and B are calibration concentrations, while in rows C-H are exposure UV fluences.**

| | 1 | 2 | 3 | 4 | 5 | 6 | 7 | 8 | 9 | 10 | 11 | 12 |
|---|---|---|---|---|---|---|---|---|---|---|---|---|
| A (Calibration) | 1.E-08 | 5.E-09 | 1.E-09 | 8.E-10 | 5.E-10 | 2.E-10 | 1.E-10 | 8.E-11 | 5.E-11 | 1.E-11 | DI | DI |
| B (Calibration) | 1.E-08 | 5.E-09 | 1.E-09 | 8.E-10 | 5.E-10 | 2.E-10 | 1.E-10 | 8.E-11 | 5.E-11 | 1.E-11 | DI | DI |
| C (LP + 5 mg $L^{-1}$ $H_2O_2$) | 0 | 50 | 100 | 150 | 200 | 250 | 300 | 350 | 400 | 500 | 600 | 700 |
| D (LP + 5 mg $L^{-1}$ $H_2O_2$) | 0 | 50 | 100 | 150 | 200 | 250 | 300 | 350 | 400 | 500 | 600 | 700 |
| E (MP + 5 mg $L^{-1}$ $H_2O_2$) | 0 | 50 | 100 | 150 | 200 | 250 | 300 | 350 | 400 | 500 | 600 | 700 |
| F (MP + 5 mg $L^{-1}$ $H_2O_2$) | 0 | 50 | 100 | 150 | 200 | 250 | 300 | 350 | 400 | 500 | 600 | 700 |
| G (MP + 10 mg $L^{-1}$ $H_2O_2$) | 0 | 50 | 100 | 150 | 200 | 250 | 300 | 350 | 400 | 500 | 600 | 700 |
| H (MP + 10 mg $L^{-1}$ $H_2O_2$) | 0 | 50 | 100 | 150 | 200 | 250 | 300 | 350 | 400 | 500 | 600 | 700 |

## Table 5.3

**Water Quality and EE2 calibration parameters for DI Water, Water 1, and Water 2.**

| | DI Water | Water 1 | Water 2 |
|---|---|---|---|
| | **Water Quality Parameters** | | |
| DOC (mg $L^{-1}$) | n/a | 3.06 | 1.39 |
| Alkalinity (mM) | n/a | 2.63 | 0.67 |
| pH | 7 | 8 | 7.98 |
| $UV_{254}$ ($cm^{-1}$) | 0.0016 | 0.04 | 0.11 |
| | **[EE2] Calibration Parameters** | | |
| B | 0.04213 | 0.04308 | 0.03997 |
| T | 1.026 | 1.027 | 1.037 |
| log[EC50] | -8.823 | -8.848 | -8.822 |
| Hillslope | 2.05 | 2.268 | 1.813 |

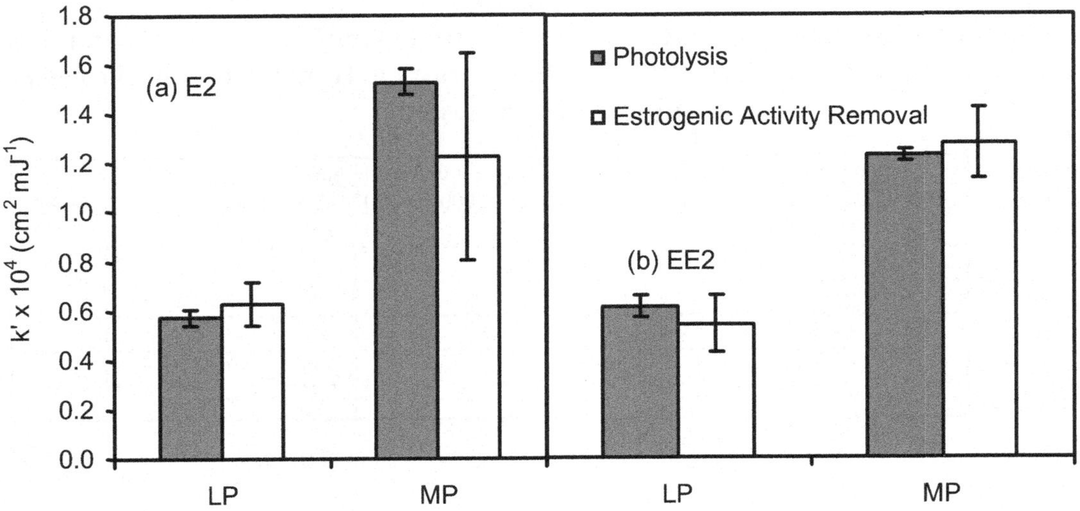


**Figure 5.1 Comparison of EDC photolysis rate and rate of estrogenic activity removal for E2 (a) and EE2 (b). Initial concentration of EE2 is and of E2 is ~5μM. Error bars represent 95% confidence intervals.**

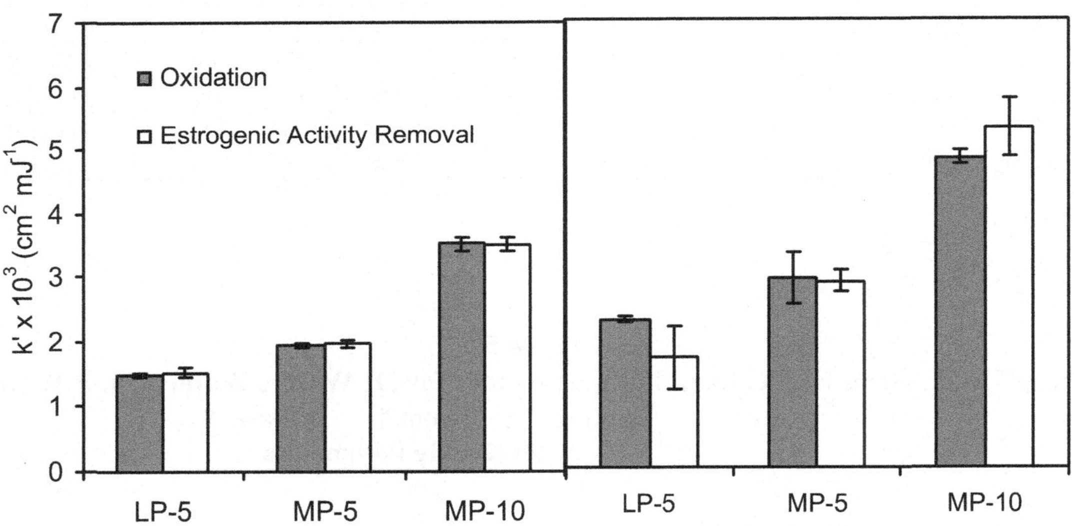


**Figure 5.2 Comparison of EDC oxidation rate and rate of estrogenic activity removal for E2 and EE2. Initial concentration of EE2 is 6.6 μM and of E2 is 7.3 μM. Error bars represent 95% confidence intervals.**

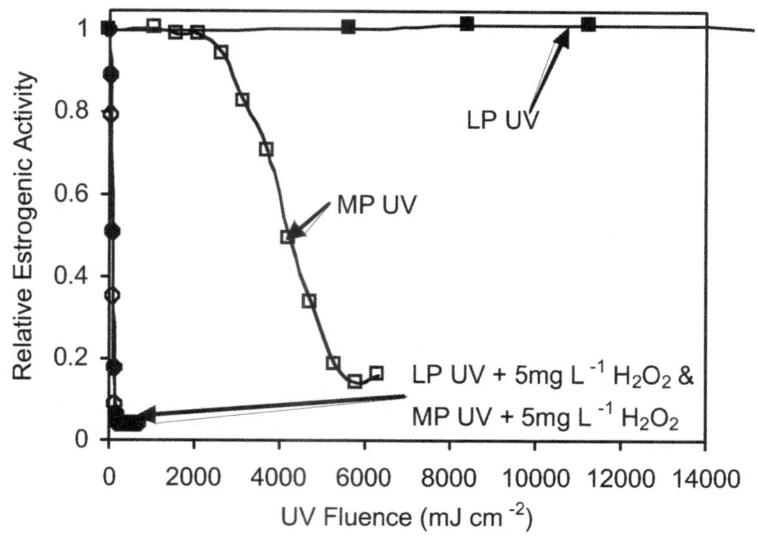


**Figure 5.3 Removal of estrogenic activity associated with 3 μg $L^{-1}$ EE2 using direct UV photolysis and UV/$H_2O_2$ AOP as a function of applied UV fluence.**

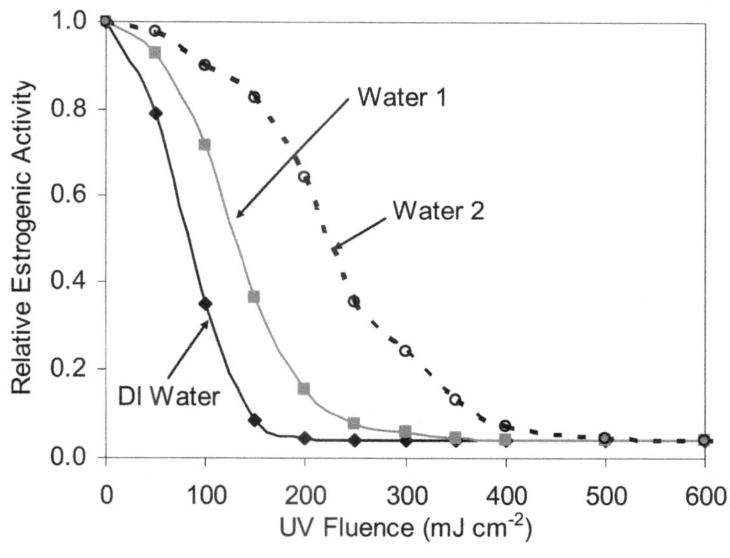


**Figure 5.4 Reduction of Estrogenic activity using MP UV + 5 mg $L^{-1}$ $H_2O_2$ AOP to oxidize 3 μg $L^{-1}$ EE2 in DI water and 2 natural waters.**

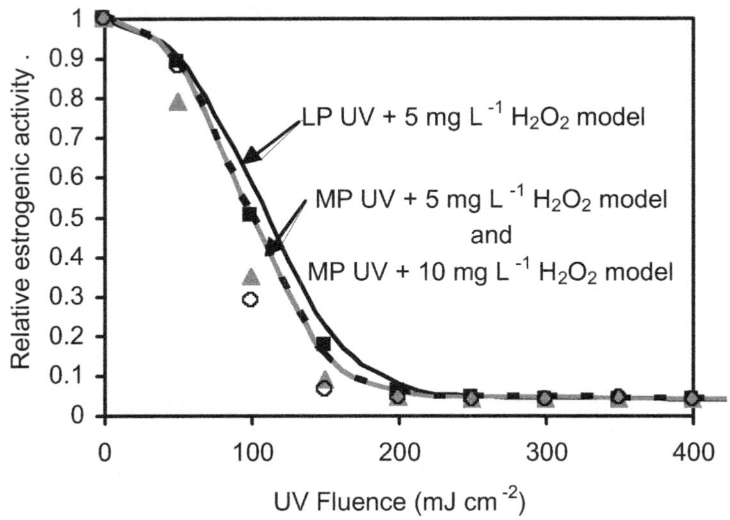


**Figure 5.5 Steady State OH radical model describing removal of estrogenic activity in DI water. Solid black squares represent LP UV – 5 data, solid grey triangles represent MP UV – 5 data, and hollow black circles represent MP UV – 10 data.**

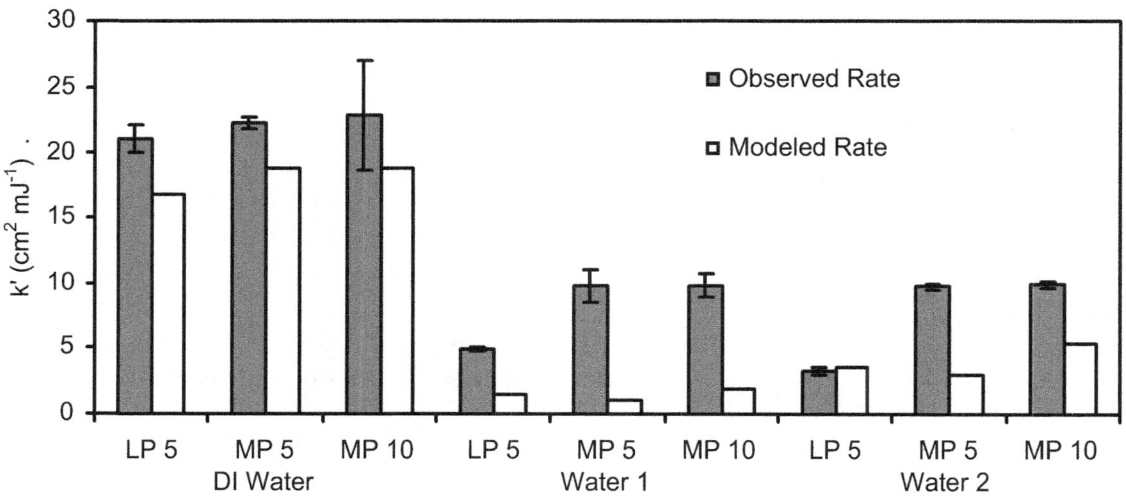


**Figure 5.6 Comparison of predicted and observed estrogenic activity removal rates associated with EE2.**

# CHAPTER 6
# WATER QUALITY IMPACTS ON BIOANAYTICAL TECHNIQUES

## INTRODUCTION

As noted above, several studies have demonstrated that the occurrence of EDCs in the environment results in adverse effects on reproduction and development in wildlife, and may reveal a significant threat to human health (Cooper and Kavlock, 1997). The U. S. Geological Survey (USGS) reported that some of the most prevalent organic wastewater contaminants in impacted US streams are detergent metabolites, plasticizers, and pharmaceuticals including potent synthetic hormones (Kolpin et al., 2002). Among hundreds of contaminants detected, nonylphenol (NP, in 50% of all samples), bisphenol-A (BPA, in 41% of all samples), 17-α-ethynyl estradiol ($EE_2$, in 16% of all samples) and 17-β-estradiol ($E_2$, in 11% of all samples) were highly present in U.S. rivers surveyed (Kolpin et al., 2002). Natural and synthetic steroids, 17-β-$E_2$ and 17α-$EE_2$ have frequently been measured in waters receiving domestic effluents. These two steroids bind to the estrogen receptor(s) and have been shown to elicit a range of physiological effects in organisms at very low concentrations (Christiansen et al., 2002). BPA and NP are synthetic alkylphenols with relatively weaker estrogenic activity, and ubiquitously distributed in the environment (Staples et al., 1998). The maximum detected concentrations of 17α-$EE_2$ (1 $\mu g\ L^{-1}$) and 17β-$E_2$ (0.2 $\mu g\ L^{-1}$) in U.S. rivers (Kolpin et al., 2002) has been shown to be capable of inducing *in vitro* transcriptional estrogenic activity in a yeast estrogen receptor screen (YES assay); also, each individual estrogen at these concentrations can elicit a range of *in vivo* estrogenic responses such as vitellogenin (VTG) production in fish following a short term exposure for each individual compound. In contrast, concentrations of individual BPA (12 $\mu g\ L^{-1}$) and NP (40 $\mu g\ L^{-1}$) detected in rivers are usually below the non-observed-effect concentration (NOEC) by these selected bioassays (Routledge and Sumpter, 1996; Schultz et al., 1998; Andersen et al., 1999; Van den Belt et al 2004). However, weak xenoestrogens (e.g., BPA and NP) in the aquatic environment may contribute certain levels of additive or synergistic estrogenic effects even though they are present below NOEC in an EDC mixture. Silva et al (2002) demonstrated that several weak estrogenic chemicals combined at concentrations below non-observed-effected concentration (NOEC) produced significant mixture effects of *in vitro* estrogenic activity in the YES assay. Also, Thorpe et al (2003) demonstrated the combined estrogenic effects of mixed steroidal estrogens in fish. However, few research evaluations on the removal efficiency of various water treatment techniques for a mixture of environmentally relevant EDCs exist in the scientific or trade literature.

Several challenges exist in the remediation of EDC- contaminated water with UV/AOPs; however, biological assessments provide a unique approach to theses challenges. First, EDCs of different potency are usually present in mixtures of compounds in the aquatic environment. Chemical based analytical methods cannot estimate the interaction or combined activities of mixed estrogenic compounds. In addition, following any oxidation process, multiple unknown transformation products may be formed. Because the estrogenic activity of most EDC degradation products is unknown, using biological assessments to evaluate the total estrogenic activity of an oxidized solution (including residual parent compounds and unknown degradation products) is an ideal way to monitor for an effective treatment endpoint.

Second, analytical studies have shown that UV/$H_2O_2$ AOPs followed first-order degradation kinetics for EDCs in the concentration range of mg $L^{-1}$ to $\mu g\ L^{-1}$ (Rosenfeldt and

Linden, 2004). Degradation kinetics for environmentally present EDCs, usually present at μg $L^{-1}$-ng $L^{-1}$ range, is difficult to evaluate due to the detection limits of analytical methods and the uncertainties that occur during extraction methods. Third, variations in natural water matrices such as UV absorbance and the presence of OH radical scavengers such as natural organic matter and alkalinity result in light screening OH scavenging variations both leading to variations in the effectiveness of destroying target compounds (Schwarzenbach et al., 1993). In addition, the natural water matrices may change the formulation of EDC-oxidation products, which in turn may lead to different levels of biological activity depending on the variability in the water quality characteristics. Because bioassays embody high sensitivity for detecting estrogenic activity of EDCs, they are capable of determining removal kinetics for transformation-based treatment processes such as the UV/$H_2O_2$ AOP of EDCs at environmentally relevant levels (μg $L^{-1}$-ng $L^{-1}$).

The objective of this research phase was to compare the efficiency of UV photolysis with the UV/$H_2O_2$ AOP for the removal of estrogenic activity of four EDCs at μg $L^{-1}$-ng $L^{-1}$ levels spiked individually or as a mixture in either laboratory deionized (DI) water or natural river water (Eno River). *In vitro* Yeast estrogen screen (YES) and *in vivo* Vitellogenin (VTG) bioassays were used to assess the reduction of estrogenic activity for different EDC- solutions following UV/$H_2O_2$ oxidation process and the interference of water matrices on UV/$H_2O_2$ oxidation were evaluated.

## MATERIAL AND METHODS

### Standards and Reagents

17-β-estradiol ($E_2$) and 17-α-ethinyl estradiol ($EE_2$) were purchased from Steraloids, Inc. (Newport, RI). Bisphenol A (BPA) and nonylphenol (NP) were purchased from Aldrich Chemicals (Milwaukee, WI). The chemical structure and properties of four EDC compounds are shown in Table 6.1. Master solutions (1 mM) of four selected EDCs were made in 95% ethanol (Sigma, St. Louirs, MO). Four EDC solutions were made from the master solutions in deionized (DI) water or natural river (Eno River) water, as concentrations given in Table 6.2, for UV/ $H_2O_2$ AOP treatments. The spiking concentration of each EDC was at the environmentally relevant level according to Kolpin, et al (2002). Each of these concentrations was within the solubility limit for each compound. The final solvent concentration in the solution was << 0.1% (v/v)/ Stocks and dilutions were stored in amber bottles at 4 $^{0}$C until use. All additional chemicals/reagents used during the experiments were of analytical grade and shown to have no impact on either bioanalytical measurements or UV/AOP processes.

### Water Quality Measurements

Eno River water was collected near Penny's Bend Nature Preserve in Durham, North Carolina. The water was collected in acetone rinsed amber glass bottles and returned to the lab where it was immediately filtered through a 0.45 μm cellulose nitrate filter (Whatman, Maidstone, Kent, UK), and stored at 4 $^{0}$C until use. Prior to chemical spiking, River and DI water were characterized by measuring total organic carbon (TOC), turbidity, alkalinity, hardness, pH, electronic conductivity (EC), $NH_3$-N and $NO_3$-N. All measurements of water quality measurements were performed in duplicate or triplicate. The results are shown in Table 6.3.

## UV/$H_2O_2$ Oxidation Experimental Design

Similar to experiments reported in previous chapters, the UV photolysis and UV/$H_2O_2$ AOP experiments were performed in a bench-scale collimated beam UV apparatus consisting of four 15 W low pressure (LP) Hg lamps (ozone-free, General Electric #G15T8), which emit monochromatic irradiation at 253.7 nm. The UV fluence (mJ $cm^{-2}$) was determined as the product of the average irradiance and exposure time based on calculations outlined in Bolton and Linden (2003). For samples used in yeast estrogen screen (YES) analysis, 50 mL of four EDC solutions was placed in a 70 × 50 mm flat bottom glass Petri dishes and spiked with 10 ppm $H_2O_2$, whose concentration was made from a 30% stock. A duplicate sample prior to treatment was taken for a water absorption scan and control YES analyses. The solution was then exposed to UV at fluences ranging from 0-2000 mJ $cm^{-2}$ at room temperature. Temperature remained constant throughout the duration of the exposure. Following UV irradiation, Catalase (Sigma) at 0.2 ppm was added to quench residual $H_2O_2$ (Liu, et al., 2003) of each sample, and then absence of $H_2O_2$ in the treated solution was verified by the Ghormley method (Klassen, 1994) after 10 minutes of stirring . A $H_2O_2$ control consisting of 50 mL of each EDC solution spiked with 10 ppm $H_2O_2$ was kept in the dark for the duration of the study. A blank control of 50 mL of Milli-Q water spiked with 10 ppm $H_2O_2$ only was exposed under the same UV/AOP protocol as other testing samples. UV/$H_2O_2$ AOP exposures were performed in triplicate for each water sample. All samples were subsequently stored at -20 $^{0}$C in 50 mL of Falcon tubes (VWR, West Chester, PA). All treated samples were freeze-dried and resuspended in DI water as 100 fold extracts for the YES assay. For VTG assays, UV/AOP exposures were conducted in 1 L of solution containing EDC's and 10 ppm- $H_2O_2$. Each experimental solution was placed in a 150 × 75 mm glass beaker and exposed to UV irradiation between 0-2000 mJ $cm^{-2}$. One liter of deionized water spiked with 10 ppm-$H_2O_2$ following UV exposure at 2000 mJ $cm^{-2}$ was used as a blank control. The EDC concentration in the dark control did not change significantly during experiments (Chen et al., 2006). After treatments, remaining $H_2O_2$ in the solution was quenched by addition of 0.2 ppm catalase and 10 minutes of stirring for the treated solutions. All treated solutions were subsequently stored 4 $^{0}$C until use.

## Yeast Estrogen Screen (YES) Assay for Treated Water

Yeast cells were provided by Dr. Sumpter at Brunel University in the U.K. The YES assay was performed as previously described in Chapters 2 and 4. Briefly, a series of 12 dilutions (1:1) were prepared in DI water from the 100- fold extract of each UV/AOP post-treated solution. Of each dilution, 100 μL were added to a 96-well deep-well microtiter plate (1 mL capacity/well, VWR) containing 300 μL of diluted yeast solutions. Besides these samples, each plate contained wells representing a standard curve with $E_2$ in DI water (final concentration dissolved in assay medium: 2.5 x $10^{-8}$ – $1x10^{-11}$ M) and a row of blanks. The color development of ortho-nitrophenyl-beta-D-galactopyranoside substrate was measured at $OD_{420}$ after incubating the microtiter plate at 30°C for 72 hr.

**Vitellogenin (VTG) Assay for Treated Water**

To perform Vitellogenin (VTG) assays for UV treated samples, four male fish (six month old) per treatment were placed in 1 L beakers containing 0.5 L of treated or non-treated EDC-solution as well as a blank control for one week. Four salts (NaCl, $CaCl_2$, KCl and $MgSO_4$) (Yamamoto, 1975) were added into each sample to ensure proper health and well being of fish. Details of water exposure, blood/serum collection and the assessment of VTG protein were as previously described in Chapters 2 and 4. Fish were feed with brine shrimp (*Artemia sp.*) daily. Half the volume of exposure solution was renewed with fresh solutions every day. After 7 days of exposure, fish blood was collected individually from the caudal vein with glass capillary tubes, and the volume was measured. Following centrifugation, serum were collected, and then combined with protease inhibitors aprotinin (0.2 mg $mL^{-1}$, Sigma) and phenylmethylsulfonylfluoride (0.5 mM, Sigma), and stored at -80 $^{0}C$ until use. VTG protein concentration in each serum sample was measured by Western blotting or numerically quantified with Medaka Vitellogenin Enzyme-Linked Immunosorbent Assay (ELISA) (Biosense Laboratories AS, Bergen, Norway).

**Data Analysis**

The sigmoid concentration-response curve of the $E_2$ standard with YES assays was fitted to a logistic function (based on Hill equation) using software Prism (GraphPad, San Diego, USA). The response of the standards and samples was expressed as percent of maximum response evoked by $E_2$. Effective concentrations to half maximum response ($EC_{50}$) for each EDC- standard were fitted with the software. The relative estrogenic activity of each EDC standard was expressed as 17β-estradiol equivalent factor (EEF), equal to the ratio of $EC_{50}$ for $E_2$ to $EC_{50}$ of the testing compounds. For each water samples (pre- and post- $UV/H_2O_2$ treatment), dilution factors to 50% of response each sample were fitted again with the software. The estrogenic activity of each water sample was expressed as 17β-estradiol equivalency concentration (EEQs), the ratio of $EC_{50}$ for $E_2$ to dilution factor for 50% response of each sample (Huber et al., 2004). 95 % confidence interval (CI) for each EEF or EEQ was computed with the Prism software. .

The $UV/H_2O_2$ AOP reaction involves the formation of the highly reactive •OH, which quickly oxidizes the EDCs as described in Rosenfeldt and Linden, 2004. In pure water, •OH scavengers include parent contaminant and their degradation products; however, in natural waters, additional •OH scavengers include natural organic matter and $HCO_3^-/CO_3^{2-}$ (Schwarzenbach et al., 1993). Scavengers and their corresponding •OH rate constants used in the model included $E_2$ (1.41 x 1010 $M^{-1}$ $s^{-1}$), $EE_2$ (1.02 x $10^{10}$ $M^{-1}$ $s^{-1}$), BPA (1.08 x $10^{10}$ $M^{-1}$ $s^{-1}$), NP (5.65 X $10^9$ $M^{-1}$ $s^{-1}$) (see Chapter 3), organic carbon (2.5 x $10^4$ L $mg^{-1}$ $s^{-1}$) (Larson, et al., 1988), $HCO_3^-$ (8.5 x$10^6$ $M^{-1}$ $s^{-1}$), $CO_3^{2-}$ (3.9 x $10^8$ $M^{-1}$ $s^{-1}$), and $H_2O_2$ (2.7 x $10^7$ $M^{-1}$ $s^{-1}$) (Buxton, et al., 1988).

## RESULTS AND DISCUSSIONS

### Combination Effects of EDC Mixtures

*In vitro* estrogenic activity (measured by YES assays) of each EDC- solution before or after UV/$H_2O_2$ treatments was quantitatively evaluated as $E_2$ equivalent concentrations (EEQs). Figure 6.1a shows the *in vitro* estrogenic activity prior to UV/$H_2O_2$ treatments for each of four EDC- solutions in deionized (DI) water or Eno River water. Model EEQ values of these four EDC- solutions are also illustrated in Figure 6.1a. The model EEQ value is the sum of EEF of each EDC times its spiking concentration. Results show that in DI water system, additive (solution 3 ~ equal to the sum of solutions 1 and 2) and synergistic (solution 4 > that the sum of $E_2$, EE, NP and BPA.) effects of *in vitro* estrogenic activity were found in EDC- mixtures, compared to the single compound solutions. However, model EEQ values of these solutions only predict the additive effects in the EDC mixtures in DI water. It is important to note here that $E_2$ and EE are the primary compounds driving estrogenic activity and that the concentrations of NP and BPA used in this study have a negligible effect on either *in vitro* or *in vivo* estrogenic activity. Interestingly, it appeared that the synergistic effect in EDC mixtures (solution 3 and 4) was more significant in Eno River water than that observed in DI water. This may be due to the presence of endogenous estrogens such as phytoestrogens (naturally occurring plant- or fungal metabolite-derived estrogen) in natural water. Endogenous estrogen in natural water may result in higher measurements estrogenicity (~2.5 times) than that of DI water (which were shown as control values of Eno River water versus that of DI water in Figure 6.1a). Furthermore, Figure 6.1b shows a putative additive and synergistic effect of *in vivo* estrogenic activity in EDC mixtures (compared to a single compound solution) in DI water following seven days of exposure in medaka male fish. The similarity of these effects in DI water suggests that mixtures of compounds with estrogenic activity may exhibit behaviors inconsistent with that observed for individual compounds. The fact that additive and synergistic effects are observed both *in vitro* and *in vivo* indicates that this is not a phenomenon driven by the mechanics of the assay used but rather chemical interactions. Given that the *in vitro* assay is specifically ER driven it is highly possible that these effects are mediated by enhancement of ER directed gene transcription. This can include alsteric interactions of the ER itself, enhanced recruitment of necessary receptor cofactors or enhanced ER-DNA binding. The exact mechanism however is not a focus of this study and to date has not been elucidated.

### Removal of *in vitro* Estrogenic Activity of EDC mixtures

UV/$H_2O_2$ oxidation of EDC solutions exhibited pseudo-first order reaction kinetics for removal of *in vitro* estrogenic activity both in DI and Eno River water, with the slope of the regression line indicating the fluence based rate constant ($mJ^{-1}$ $cm^2$), *k'*, as shown in Figure 6.2 and 6.4. In addition, Figure 6.2 illustrates that UV in combination with 10 ppm of $H_2O_2$ at UV fluence levels relevant for UV oxidation (<1000 mJ $cm^{-2}$) significantly decreased *in vitro* estrogenic activity of EDC solutions at environmentally relevant concentrations. However, the removal rate of *in vitro* estrogenic activity for an EDC- mixture (solution 4) was lower than that observed in the solution with a single compound. This finding was consistent both in DI water (Figure 6.2a) and Eno River water (Figure 6.2b) systems. In this study, Eno River water contained 0.89 mg $L^{-1}$ TOC, 4.2 NTU of turbidity and slight hardness, conductivity, nitrate and chlorine (Table 6.3).

**Removal of *in vivo* Estrogenic Activity of a Mixture of EDCs**

Figure 6.3a shows Western blotting analysis of VTG induction in medaka fish after 7 days of exposure to solution 4 (a mixture of four EDCs) following UV + 10-ppm $H_2O_2$ oxidation treatment between 0-2000 mJ $cm^{-2}$ of UV irradiation. As seen in the figure, VTG induction decreased with UV fluence, resulting in non-detectable VTG induction after at 2000 mJ $cm^{-2}$ of UV irradiation in the presence of 10 ppm $H_2O_2$ in DI water. However, in Eno River water spiked with solution 4, VTG induction was prolonged following identical treatment conditions. This result suggests that UV+ 10- ppm of $H_2O_2$ at UV fluence relevant to water treatment (<1000 mJ $cm^{-2}$) was not sufficient to completely remove *in vivo* estrogenic activity of this EDC mixture. Figure 6.3b demonstrates the ELISA quantification of VTG protein induction in medaka fish after seven days of exposure to solution 4 following UV + 10-ppm $H_2O_2$ oxidation treatment in both DI and Eno river water. First or pseudo-first order kinetics of *in vivo* estrogenic activity was followed with this treatment design in both DI and Eno River waters. However, it appeared that the removal rate of *in vivo* estrogenic activity in DI was higher than that was observed in natural river water.

**Comparisons of Removal Rates**

Figure 6.4 compares removal rate constants of four EDC- solutions following UV + 10 ppm $H_2O_2$ AOP in DI and natural waters using both between *in vivo* and *in vitro* assays. The removal rate of *in vitro* estrogenic activity for EDC- mixtures was significantly lower than that observed in the solution with a single compound. In addition, for both *in vitro* and *in vivo* estrogenic activities removal rates were slightly faster in DI water than those observed in natural water with UV/$H_2O_2$ AOP. One explanation for the slower removal rates in natural water is a decrease in the steady state •OH concentration due to the scavenging effect of the dominant estrogenic species, $E_2$ and $EE_2$, and natural water components such as TOC. The overall effects would manifest as a reduced the degradative capacity and slower rate constant. Table 6.4 lists the distribution of •OH radicals reacting with each •OH scavenging species in the solutions. Results demonstrate that in natural water, total organic carbon (TOC) consumed $\geq$70% of the •OH generated in natural water. In DI water, slightly less •OH reacted with two estrogens in the mixture (solution 4), compared with the ratio of •OH in solution 1 ($E_2$ alone) or 2 ($EE_2$ alone). This is most likely due to the presence of high concentrations of BPA and NP in the mixture, which also contributed to the •OH demand, consuming a total of 16.6 % of •OH. The reduction in the removal rate of estrogenic activity was likely due to a decreased steady-state concentration of •OH in the presence of these scavengers (e. g. high concentration of BPA and NP in DI water and TOC in natural water), and thus slowed •OH reactions for the degradation of $E_2$ and $EE_2$. Other explanations include that water quality parameters such as TOC or turbidity in the Eno River water antagonize the $EE_2$- estrogenic activity *in vitro* and *in vivo* with YES and VTG bioassays. In addition, Figure 6.4 shows that removal rate of *in vivo* estrogenic activity from a mixture of EDCs (solution 4) was faster than that measured *in vitro* following UV+10 ppm-$H_2O_2$ AOPs. This is likely explained by differences in pharmacokinetic properties between yeast cells which do not metabolize E2, EE, NP or BPA and whole fish which may sequester these compounds rendering them less bioavailable.

## CONCLUSIONS

Our results demonstrated that combined (additive or synergistic) effects of *in vitro* and *in vivo* estrogenic activity were found in EDC mixtures compared to a single compound in solution. In addition, it appeared that the synergistic effect of *in vitro* estrogenic activity in EDC mixtures was more significant in Eno River water than that observed in DI water due to the possible presence of endogenous estrogens in natural water. Our results suggest that UV/$H_2O_2$ AOP exhibited first-order kinetics for removing both *in vitro* and *in vivo* estrogenic activity for a mixture of EDCs at environmentally relevant concentrations ($\mu g\ L^{-1}$-$ng\ L^{-1}$) in DI and natural water systems. UV/$H_2O_2$ treatment at $H_2O_2$ doses (10 ppm) and UV fluence (<1000 mJ $cm^{-2}$) relevant to water treatment used for chemical oxidation processes was capable of decreasing *in vitro* and *in vivo* estrogenic activity for an EDC mixture at low concentrations. In addition, removal rates of *in vitro* estrogenic activity of EDC mixtures was lower than that observed with single compounds alone. Also, removal of *in vitro* estrogenic activity with UV/$H_2O_2$ AOPs was slightly slower in natural water, compared with that observed in laboratory water. The reduction in the removal rate of estrogenic activity is likely due to a decrease in the steady-state concentration of •OH due to the presence of scavengers and thus a reduced rate of reaction for the degradation of $E_2$ and $EE_2$. Although removal rate of *in vivo* estrogenic activity was faster than that of *in vitro* for a mixture of EDCs with the treatment of UV+ 10-ppm of $H_2O_2$, a UV fluence of 2000 mJ $cm^{-2}$ was not efficient enough to completely remove *in vivo* estrogenic activity in natural water.

From this research, we have demonstrated that the use of biological assessments provides a unique approach to evaluate the performance of UV/$H_2O_2$ AOPs for $\mu g$-$ng\ L^{-1}$ concentrations of EDC contaminated water. In conjunction with standard analytical approaches this assessment design will enable estimates of the interactions and/or combined estrogenic activities among mixtures of EDCs and water matrices in natural water. In addition, bio-assessment of remaining estrogenic activity following UV treatment demonstrated differences between *in vitro* and *in vivo* analyses. These differences highlight the differential sensitivities of *in vitro* and *in vivo* bioanalytical assays and reemphasize the necessity to employ mechanistically diverse assays for assessment of EDC- activity in treatment studies.

**Table 6.1**
**Chemical and physical properties of selected EDCs**

| Compounds | Estradiol ($E_2$) | Ethynyl estradiol ($EE_2$) | Bisphenol A (BPA) | Nonylphenol (NP) |
|---|---|---|---|---|
| Chemical Structure | 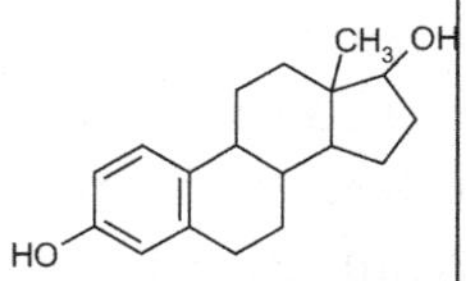 | 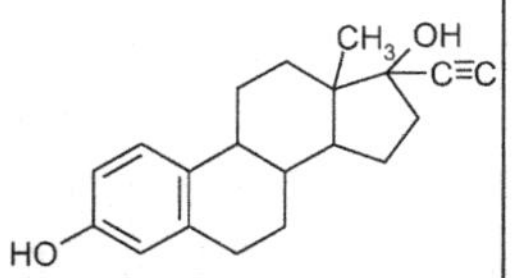 | 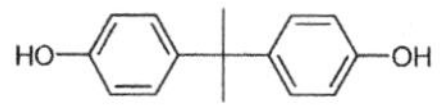 | 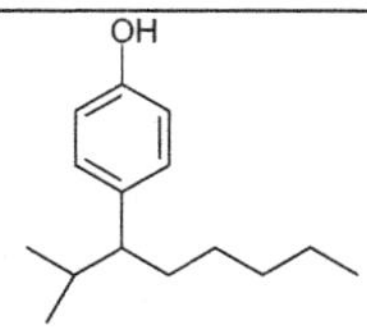 |
| Water solubility [a] | 3.6 mg $L^{-1}$ at 27 °C | 11.3 mg $L^{-1}$ at 27 °C | 129 mg $L^{-1}$ at 25 °C | 4.9 mg $L^{-1}$ at 25 °C [c] |
| Log Kow | 4.01 | 3.67 | 3.32 | 4.48 [c] |
| Max-Med conc. in US surface water [b] (ng/L / μM) | 200-9/ (73-3.3)*E-05 | 831-73/ (28-2.5)*E-04 | 12000-140/ (526-6.1)*E-03 | 40000-800/ (182-3.6)*E-03 |
| $EC_{50}$ in YES (μM) | 3.40*E-04 | 3.02*E-04 | 7.23 | 2.20 |

[a] Physical data from Chemfinder.com.

[b] Maximum-Medium concentration of EDCs in US river water measured by chemical analysis (Kolpin et al., 2002).

[c] Brix et al., 2001.

**Table 6.2**
**Experimental design of EDC-solutions for UV/$H_2O_2$ AOP treatments**

| **Spiking conc. of EDCs[a]** | **$E_2$ (ng $L^{-1}$)** | **$EE_2$ (ng $L^{-1}$)** | **BPA (ng $L^{-1}$)** | **NP (ng $L^{-1}$)** |
|---|---|---|---|---|
| Solution 1 ($E_2$) | 200 | - | - | - |
| Solution 2 ($EE_2$) | - | 1000 | - | - |
| Solution 3 ($E_2$+$EE_2$) | 200 | 1000 | - | - |
| Solution 4 ($E_2$+$EE_2$+BPA+NP) | 200 | 1000 | 12000 | 40000 |

[a] The spiking concentration of each EDC was at environmentally relevant levels based on Kolpin et al. (2002).

**Table 6.3**
**Water quality parameters of selected water (prior to addition of EDCs)**

| Water sources | TOC mg $L^{-1}$ | | Turbidity NTU | | Alkalinity mg $L^{-1}$ $CaCO_3$ | | Hardness mg $L^{-1}$ $CaCO_3$ | | pH | | EC µS | | Nitrate mg $L^{-1}$ | Total chlorine mg $L^{-1}$ |
|---|---|---|---|---|---|---|---|---|---|---|---|---|---|---|
| | AVG | SD | AVG | SD | AVG | SD | AVG | SD | AVG | SD | AVG | SD | AVG of duplicate | AVG of duplicate |
| DI water (deionized milli-Q) | 0.00 | 0.00 | 0.1 | 0.0 | 3.7 | 1.8 | 0.0 | 0.0 | 6.67 | 0.09 | 0 | 0 | 0.0 | 0.00 |
| Eno water (Eno River) | 0.89 | 0.03 | 4.2 | 0.1 | 28.3 | 2.9 | 44.0 | 4.0 | 7.59 | 0.03 | 96 | 3 | 0.7 | 0.09 |

**Table 6.4**

**Distribution of •OH reacting with each OH radical scavenging species in EDC solutions**

| | DI + peroxide | | Eno River + peroxide [a] | |
|---|---|---|---|---|
| | % | M | % | M |
| **In Solution 1 [b]** | | | | |
| •OH reacting with $E_2$ | 0.13 | 3.52E-15 | 0.03 | 1.81E-16 |
| •OH reacting with $H_2O_2$ | 99.87 | 2.70E-12 | 26.24 | 1.39E-13 |
| •OH reacting with TOC | | | 73.73 | 3.90E-13 |
| [•OH]ss | 100 | 2.70E-12 | 100 | 5.29E-13 |
| **In Solution 2** | | | | |
| •OH reacting with $EE_2$ | 0.43 | 1.16E-14 | 0.11 | 6.01E-16 |
| •OH reacting with $H_2O_2$ | 99.57 | 2.68E-12 | 26.22 | 1.39E-13 |
| •OH reacting with TOC | | | 73.67 | 3.90E-13 |
| [•OH]ss | 100 | 2.69E-12 | 100 | 5.29E-13 |
| **In Solution 3** | | | | |
| •OH reacting with $E_2$ | 0.13 | 3.49E-15 | 0.03 | 1.81E-16 |
| •OH reacting with $EE_2$ | 0.43 | 1.16E-14 | 0.11 | 6.01E-16 |
| •OH reacting with $H_2O_2$ | 99.44 | 2.68E-12 | 26.21 | 1.39E-13 |
| •OH reacting with TOC | | | 73.64 | 3.89E-13 |
| [•OH]ss | 100 | 2.69E-12 | 100 | 5.29E-13 |
| **In Solution 4** | | | | |
| •OH reacting with $E_2$ | 0.11 | 2.42E-15 | 0.03 | 1.63E-16 |
| •OH reacting with $EE_2$ | 0.36 | 8.06E-15 | 0.11 | 5.42E-16 |
| •OH reacting with BPA | 5.93 | 1.33E-13 | 1.78 | 8.94E-15 |
| •OH reacting with NP | 10.71 | 2.40E-13 | 3.22 | 1.62E-14 |
| •OH reacting with $H_2O_2$ | 82.90 | 1.86E-12 | 24.90 | 1.25E-13 |
| •OH reacting with TOC | | | 69.97 | 3.51E-13 |
| [•OH]ss | 100 | 2.24E-12 | 100 | 5.02E-13 |

[a] % •OH radicals to react with $HCO_3^-/CO_3^{2-}$ is low enough to be neglected.

[b] The distribution of •OH reacting to its scavenging species (e.g. EDCs, TOCs) in each solution was calculated using the approach outlined in Chapter 3.

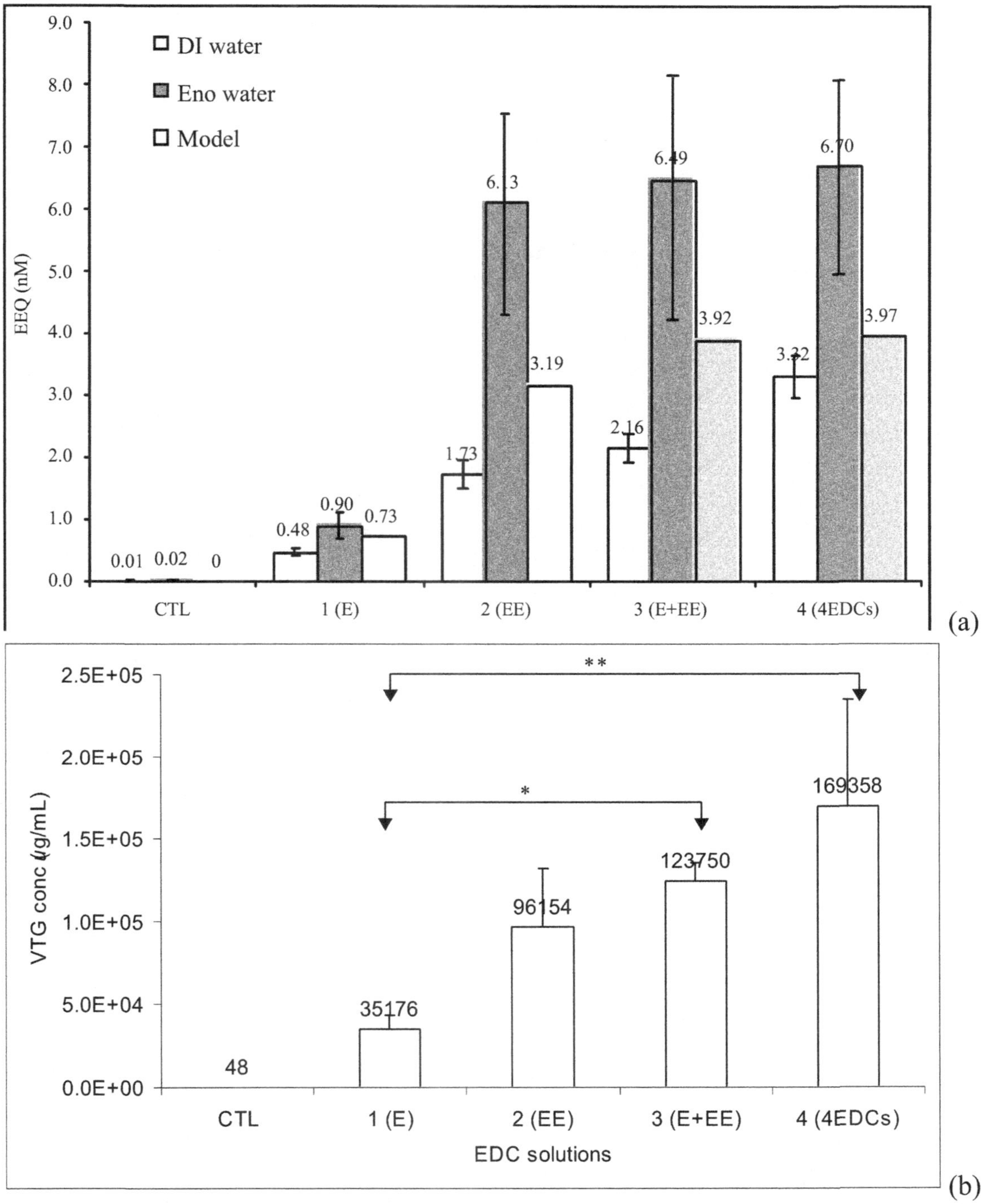


**Figure 6.1 (a) YES-estrogenic activity of four EDC- solutions in deionized (DI) water or Eno River water and model values (before UV/AOP); (b) VTG induction in medaka fish following 7 days of water exposure to four EDC-solutions in DI water (before UV/AOP). In (a), error bars represent 95% confidence intervals (CIs) of the estimate mean. In (a), the model EEQ value is the sum of EEF of each EDC in the solution times its spiking concentration. In (b), error bars represent standard deviation (SD) of the estimate mean. * (90%) and ** (95%) significant difference.**

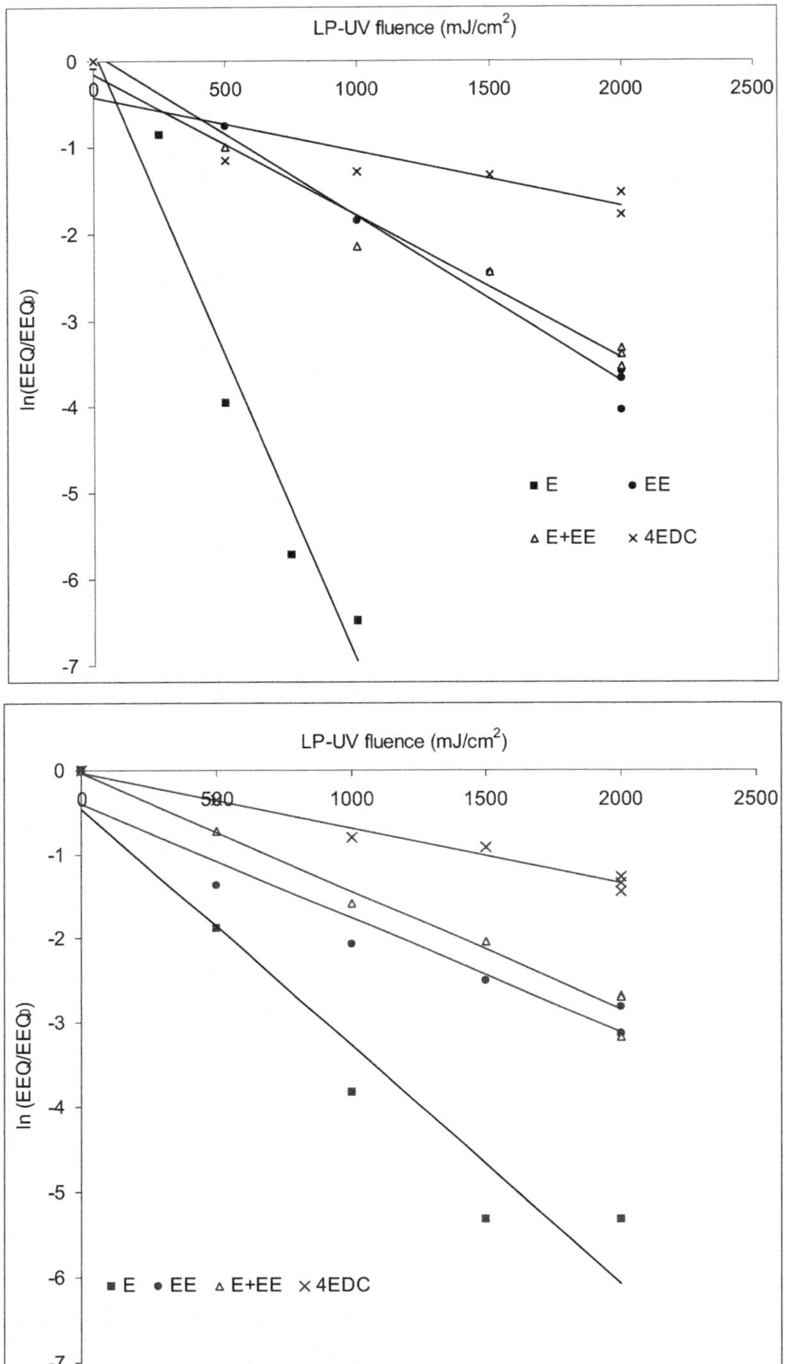


(a)



(b)

**Figure 6.2 Removal kinetics of YES- estrogenic activity for four EDC- solutions in (a) DI water or (b) Eno River water following LP- UV + 10 ppm $H_2O_2$ between 0-2000 mJ cm$^{-2}$.**

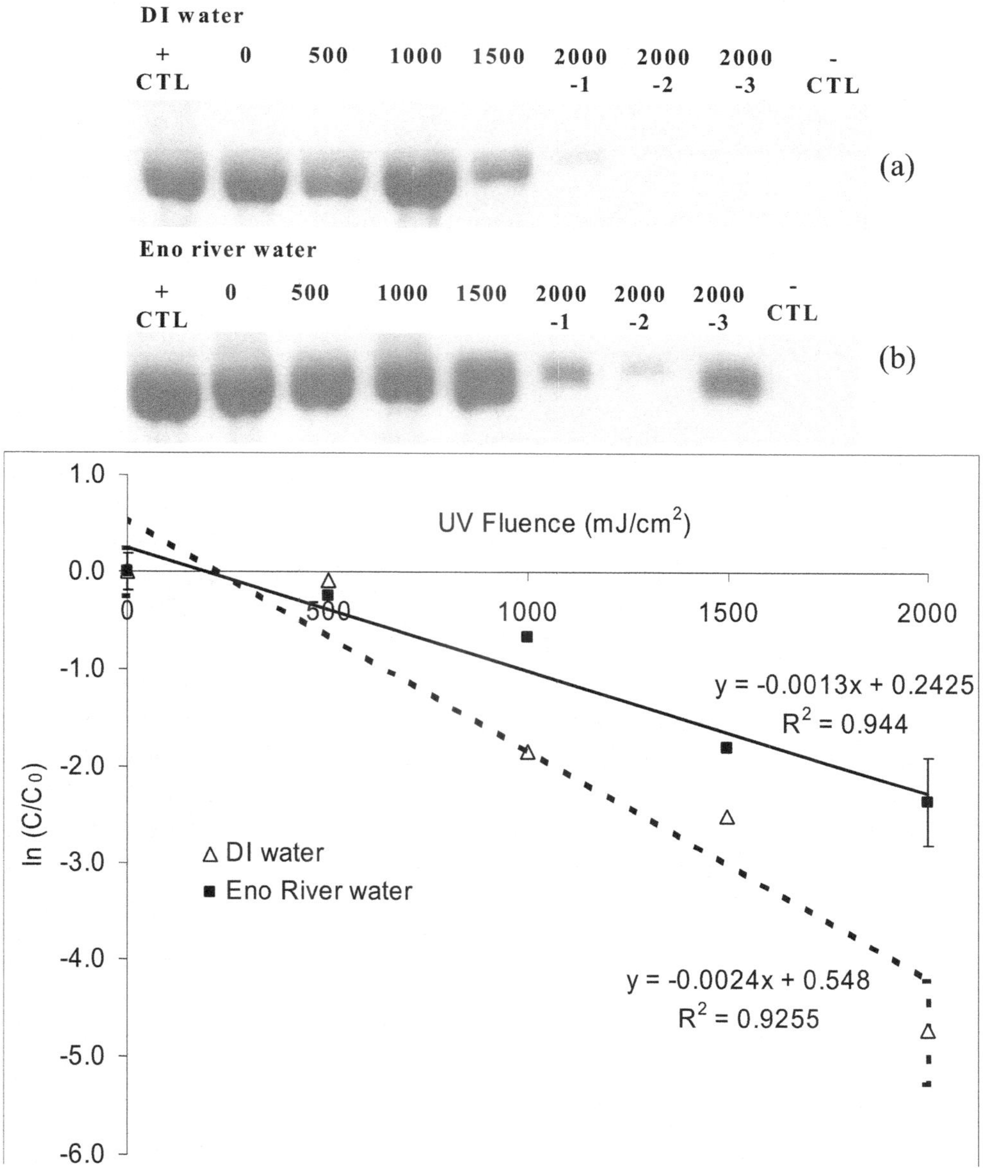


**Figure 6.3 Removal of VTG- estrogenic activity for solution 4 (with four EDC- compounds) in DI water or Eno River water following LP-UV + 10 ppm $H_2O_2$ with (a) Western analysis, or (b) ELISA quantification. Error bars represent standard deviation (SD) of the estimated mean. C represents VTG concentrations.**

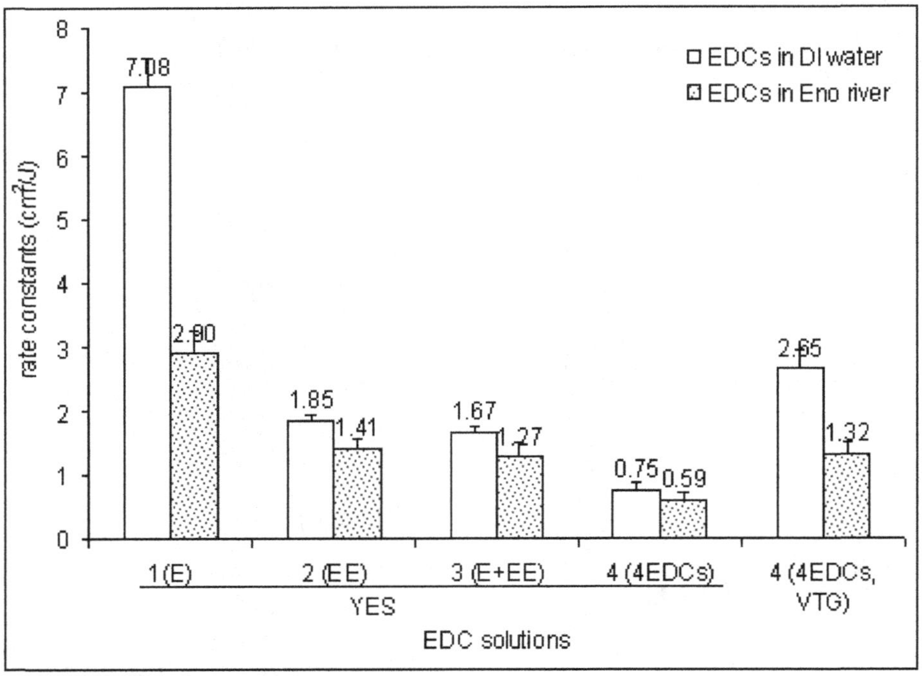


**Figure 6.4 Comparisons of removal rates between *in vitro* and *in vivo* as well as between DI and Eno River water. The treated condition was followed LP-UV + 10 ppm $H_2O_2$ between 0-2000 mJ $cm^{-2}$. Error bars represent standard deviation (SD) of the estimated mean. The difference in removal rates of estrogenic activity among solutions was not statistically significant even at 80% confidence intervals.**

# CHAPTER 7
# SURVEY OF ESTROGENIC ACTIVITY AT WATER TREATMENT PLANTS

## INTRODUCTION

Persistent EDCs appear to remain in the surface waters into which they are discharged on the order of days, weeks, or months (Ying et al.2002, Jurgens et al. 2002, Kang and Kono 2002). This is enough time for them to travel significant distances downstream, possibly infiltrating recreation areas, habitat sanctuaries, or drinking water intakes. Of particular concern to human health is the latter, because EDCs are active at very low concentrations, and conventional drinking water treatment schemes are not specifically designed to remove EDCs.

EDCs, as well as other pharmaceutical and personal care products (PPCP), manage to enter surface waters through wastewater treatment processes (natural and synthetic hormones and PPCPs) or by agricultural runoff and irrigation return waters (pesticides and animal husbandry hormones and medicines). A recent nationwide survey of pharmaceuticals in US surface water found many of these EDCs at ng $L^{-1}$ levels in 139 stream sites throughout the United States, including three sites in eastern North Carolina. (Kolpin, et al, 2002).

The water treatment plants (WTPs) in eastern North Carolina tested in this study draw source water from three main river sources, the Tar River, Neuse River, and the Cape Fear River. These rivers' headwaters are located in the relatively industrialized piedmont region of North Carolina, from the Piedmont Triad region (Winston-Salem, Greensboro, and High Point) in the west to the Research Triangle region (Raleigh, Durham, Chapel Hill) in the east. These cities draw from the rivers, and return wastewater to the rivers, impacting them as they flow toward the eastern region of the state. However, because of the point source and non-point sources of EDCs to surface waters, the agricultural impacts of the eastern region of the state may contribute EDCs to the waters as well. Table 7.1 displays relevant information regarding the tested WTPs, including their inlet water sources and treatment processes employed.

Conventional treatment was defined as initial flocculation/coagulation, settling, filtration, and either chlorination or chloramination for disinfection. Preoxidation with ozone was performed at the WTP- 6, and WTP- 2, utilities located at either end of the Cape Fear River. WTP- 3, and WTP- 1, each with conventional treatment processes, also draw from points along the Cape Fear River. WTP- 4 and WTP- 5, servicing a small city in eastern North Carolina, draw from the Tar River, while the WTP- 7 and WTP- 8 both draw from tributaries of the Neuse River.

The objectives of this phase of the research were to utilize the Yeast Estrogen Screen (YES) to examine the estrogenic activity of the raw inlet waters to each plant, as well as the effluent waters entering the distribution system from each plant. In addition, a sample along the treatment process, typically after settling but before the addition of chlorine, was examined for estrogenic activity. Recognizing the hydrophobic nature of many of these EDCs, the solids associated with the inlet waters were filtered from the water matrix, and tested for estrogenic activity.

## MATERIALS AND METHODS

### Utilities Sampling

Eight water plants were visited and sampled based on several factors, including size, proximity, and perceived risk of their source water to contamination by EDCs. The plants, and their water sources are described in Table 7.1.

Sampling visits consist of accompanying a plant manager, director, operator, or engineer on a tour of the facility, to become familiar with the treatment processes employed in the facility. After discussing the process specifics, 1 L glass bottles were filled to no headspace with samples from the plant influent (2x1L), in the process after settling or filtration but before chlorination (1x1L), and the finished effluent (1x1L). Samples were immediately labeled, packaged in bubble wrap and other packing material, and placed on ice at 4°C in a cooler. After transporting back to Duke University, samples were analyzed using the Yeast Estrogen Screen (YES) within 2 days of collection.

### Analysis of Estrogenic Activity

#### *Sample Preparation*

Each water sample was tested for estrogenic activity with the YES with and without concentration. Figure 7.1 displays a schematic plate map for the YES testing, designed to determine if the waters displayed estrogenic activity of the whole waters as is, or with a modest 50:1 or 25:1 concentration factor.

In addition, the solids were extracted from the inlet water using a 45 micron, glass fiber, TSS filter. The solids were then washed with 20 mL methanol, and the extract collected for YES analysis. After washing, the collected solids were then baked at 105°C for 1 hour as per Standard Methods, and weighed to measure Total Suspended Solids. The concentrating process allowed estrogenic activity to be detected to an estrogen equivalent concentration of $1 \times 10^{-11}$ M (~3 ng $L^{-1}$). Without concentration, the detection limit for the assay was approximately $5 \times 10^{-10}$ M (~150 ng $L^{-1}$) estrogen equivalence. The testing also examined the estrogenic activity associated with the solids in the raw water.

#### *Yeast Estrogen Screen (YES)*

The YES assay was performed according to the method by Routledge and Sumpter (1996) with modifications as described in Chapters 2 and 4. Briefly, 100 µl dilutions of BPA standards (0-100 µM) and post treated solutions were added to 96 deep-well plates (1 mL capacity/well, VWR, West Chester, PA) containing 300 µl of diluted yeast solution an optical density (O.D.) of 0.08 at 630 nm. Plates were covered with Kimwipes papers (Kimtech science, Roswell, GA) and incubated at 30°C for 3 days with gentle shaking. 17β-estradiol (0.0001- 10 nM) served as a positive control. After 3 days of incubation, 400 µl of assay buffer (containing ortho-nitrophenyl-beta-D-galactopyranoside (ONPG) substrate and lysis buffer) was then added to each well. Plates were sealed with an adhesive seals membrane (USA Scientific Inc., Ocala, FL) and incubated at 30 °C for 20 minutes. 200 µl of 1 M sodium carbonate was used to stop □-

galactosidase reactions in the media. Plates were centrifuged at 3,000 rpm for 10 minutes. To determine O.D. at 405 nm and 630 nm, 100 μl of supernatant was taken from each well and then transferred to a new 96 well microtiter plate. All values were presented as OD405 minus OD630.

***Water Quality Parameters***

TOC, alkalinity, pH and UV absorbance were measured using standard methods. Total organic carbon (TOC) was measured in accordance with Standard Method 5310 A (combustion and detection method) (Tekmar Dohrmann Apollo 9000 Total Carbon analyzer). Alkalinity was measured according to Standard Methods 2320 B. pH was measured using an electronic pH meter (Cole Parmer pH 100 series), calibrated daily using pH 4, 7, and 10 buffers. UV absorbance (200-300 nm) of the spiked test water was measured in a spectrophotometer (Cary 100 bio-spectrophotometer, Varian, Houston, TX).

## RESULTS AND DISCUSSION

### YES Activity of Unconcentrated Water Samples

Preliminary estrogenic activity results for the whole water samples and solids in the first five water plants tested are compiled in Figure 7.2. The data includes the raw inlet water, which is the water coming into each plant, and the effluent water, which is the water exiting the plant into the distribution system. Also included is a water along the process scheme, typically after the solids settling step, and before the addition of a chemical oxidant above the filters. However, in the WTP-2 water, the process water is from the effluent of the initial ozone contactor, before the addition of any settling agents.

Examining Figure 7.2 indicates negligible response in all of the water samples with the exception of WTP- 1 effluent and WTP- 7 process. Surprisingly, there was a slight estrogenic response found in these waters, findings which prompted the concentration experiments. The detection limit of the assay is 200 ng $L^{-1}$ of estrogen equivalence, indicating that if any estrogenic EDCs are present in any of the waters with minimal response, their total sum activity is not equal that of 200 ng $L^{-1}$ of 17-β-estradiol.

### YES Activity of Concentrated Samples

To further investigate the estrogenic activity of the samples, the samples were concentrated using C-18 Solid Phase Extraction (SPE) cartridges. The concentration process varied, but all samples were concentrated 50x or 25x, and eluted in methanol. Then, YES was performed diluting the samples 1:2 in each well, and a concentration factor was found which corresponded to 50% relative estrogenic activity ($CF_{50}$), where appropriate. In some cases, no positive response was observed. When a response was elicited, however, equation 7.1 was utilized to transform the $CF_{50}$ information into an estrogen equivalent concentration (EEQ).

$$EEQ = \frac{[EC_{50}]}{CF_{50}} \tag{7.1}$$

where $[EC_{50}]$ is the molar concentration of E2 eliciting a 50% relative estrogenic response in calibration experiments.

Figure 7.3 displays the maximum relative estrogenic activity response for each water sample, normalized to the maximum positive response for the estrogen calibration. In cases where the maximum estrogenic response was measurable (>80%), an EEQ value was calculated for the water, and is displayed in the inset table.

In examining Figure 7.3, the first observation to note is the inconsistencies between the results presented here, and in Figure 7.2. First, with respect to the WTP- 6 inlet water, no measurable response was observed in the whole water, indicating an estrogenic response of less than 200 ng $L^{-1}$ estrogen equivalents. However, in the concentrated sample, an estrogenic response equivalent to 750 ng $L^{-1}$ E2 was observed. Secondly, in WTP- 1 effluent water, a slight response was observed in the unconcentrated sample, but no response was observed in the concentration experiments. Reasons for these inconsistencies were not determined in this experiment, but must be examined in future research.

The concentrated water samples with measurable YES responses the inlet water to WTP- 6, and all of the water samples in WTP- 7 and WTP- 8, and the inlet and process water samples of WTP- 2. This result may be correlated to the location of the source waters for these plants. All of the plants draw from surface water sources, but the four plants which display no YES response at 50x concentration are located in more rural settings than those displaying some responses. This is a preliminary observation and must be confirmed with more testing to study these trends.

**YES Activity in the Solid Fraction**

To further investigate the solids, a mass based estrogenic equivalent activity was calculated ($EEQ_m$) as the mg of estrogen equivalence per kg of solids in each influent water. 1:2 dilutions of the solids extract were analyzed with YES, and Figure 7.4 is a plot of the relative estrogenic activity as a function of concentration factor (relative to 1 L of the influent water from which the solids were extracted).

From the data presented in Figure 7.4, the concentration factor for 50% estrogenic response ($CF_{50}$) was determined by fitting the data (relative response as a function of log(CF)) to a variable slope sigmoidal dose response curve in the statistical package Graphpad Prism. The 17-β-estradiol (E2) calibration curve accompanying each run was also input into this package (relative response as a function of log([E2]), and yielded an $EC_{50}$ corresponding to the concentration of E2 yielding 50% response. By dividing the $EC_{50}$ by $CF_{50}$ for each set of solids, and estrogen equivalent concentration (EEQ) was determined for the solids in 1 L of influent water. This data was then divided by the corresponding TSS value for the water, to produce an $EEQ_m$ (mg EEQ (kg solids)$^{-1}$). The $EEQ_m$, $CF_{50}$, and EEQ for the solids are displayed in Table 7.2.

The solids from WTP- 5 plant did not display estrogenic activity with a 100:1 concentration, so the $EEQ_m$ was reported as less than the minimum possible, and the solids from WTP- 2 were not diluted enough to observe the curve, and as such the $EEQ_m$ is reported as greater than the maximum possible. The mass of estrogen equivalents in the solids is relevant because of the hydrophobic nature of many xenoestrogenic compounds. These compounds will conglomerate around particles in water, placing emphasis on water treatment processes capable of settling or filtering particles from the water.

## Ozone Treatment for Removal of Estrogenic Activity

At both WTP- 2 and WTP- 6, the inlet water is pre-treated with ozone before entering the flocculation/ sedimentation basins. This provided a unique opportunity to examine the effect of a strong oxidant on removing the estrogenic activity within the water. Table 7.3 indicates the estrogenic activity equivalence for the aqueous and solid fractions of the raw water before and after the pre- ozonation process. In WTP- 2 water, a dose of 5 mg $L^{-1}$ is applied, while only 2 mg $L^{-1}$ ozone was applied at the WTP- 6 plant.

As this data shows, the difference between the EEQ in the water fraction is negligible, and not statistically significant in WTP- 2 water, but there is nearly a 1000 fold removal of estrogenic activity in WTP- 6 water. The difference between the treatment of these waters is the added settling process in the WTP- 6 system, but the estrogen equivalents associated with the solid fraction in the WTP- 2 water would indicate that this does not fully explain the differences. The EEQ associated with the solids fraction in the WTP- 2 water is an order of magnitude less after ozonation. This was an interesting result, given that ozone may be more likely to oxidize compounds in solution before those associated with particles. One possible explanation for this result might be equilibrium chemistry. As the estrogenic species are being oxidized in the water, there may be a concentration driving force for the estrogenic species associated with the solid fraction to desorb into the water. Another possibility may be that the oxidation by ozone is transforming non-estrogenic micropollutants in the inlet water into more estrogenic species. This has been observed when bisphenol-A (Hu et al. 2002a) and nonylphenol (Hu et al. 2002b) are oxidized with chlorine species, but it typically not observed in ozone oxidation. However, more research needs to be performed to further examine the reasons behind these observations.

## CONCLUSIONS

In general, no observed estrogenic activity as measured with the YES was observed in the unconcentrated water samples collected from the utilities studied in this project. The few samples with slight responses were most likely indicative of errors or artifacts associated with this assay. This result was further supported when the samples were concentrated and no response was observed in the concentrated sample corresponding to the WTP- 1 effluent sample. The concentrated results yielded some interesting observations, including an apparent correlation between the location of the water source within an urban to suburban area and increased estrogenic activity. Utilities with sources located in more rural or small town settings did not show positive responses to the YES. Estrogenic activity was found associated with all the solids in the inlet waters with the exception of WTP- 5. This was expected due to the hydrophobic nature of many xenoestrogenic compounds, and a mass based estrogen equivalent concentration for the solids (mg $kg^{-1}$) was calculated. Finally, pre-oxidation with ozone was determined to remove estrogenic activity associated with the solid fraction of raw water.

**Table 7.1**
**Tested Water Treatment Plants**

| Utility | Inlet Water Source | Treatment Process |
|---|---|---|
| 1 | Cape Fear River | Conventional |
| 2 | Cape Fear River | Conventional + preoxidation + biological filtration |
| 3 | Cape Fear River | Conventional |
| 4 | Tar River | Conventional |
| 5 | Tar River | Conventional |
| 6 | Cape Fear River (Lake) | Conventional + preoxidation |
| 7 | Neuse River (Lake) | Conventional |
| 8 | Neuse River (Lake) | Conventional |

**Table 7.2**
**Mass based estrogen equivalence ($EEQ_m$) for the solids in each influent**

| Utility | CF50 | EEQ (M) | Mass EEQ in solids (mg/kg) |
|---|---|---|---|
| 1 | 11.38 | 1.08E-10 | 5.7 |
| 2 | n/a | n/a | >14.4 |
| 3 | 2.0 | 6.23E-10 | 16.80 |
| 4 | 18.5 | 6.67E-11 | 4.13 |
| 5 | n/a | n/a | <1.40 |
| 6 | 56.6 | 2.18E-11 | 1.98 |
| 7 | 18.8 | 6.57E-11 | 11.2 |
| 8 | 1.6 | 7.90E-10 | 14.35 |

Table 7.3
**Estrogen Equivalent concentrations for the water fraction and solid fraction of WTP-2 and WTP-6 raw waters before and after ozone pre-oxidation**

| Sample | Water Fraction EEQ (ng $L^{-1}$) | Solids Fraction $EEQ_m$ (mg $kg^{-1}$) |
|---|---|---|
| 2 -Raw | 191 +/- 31.6 | >14.4 |
| 2 - After Ozone | 176 +/- 53.1 | 1.84 |
| 6 - Raw | 752 | 5.67 |
| 6 - After ozone + settling | 0.957 | n/a |

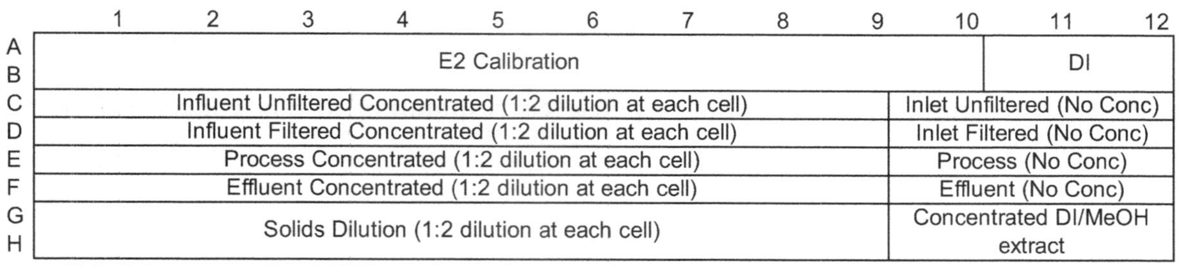


**Figure 7.1 Plate map for YES testing of water from utilities**

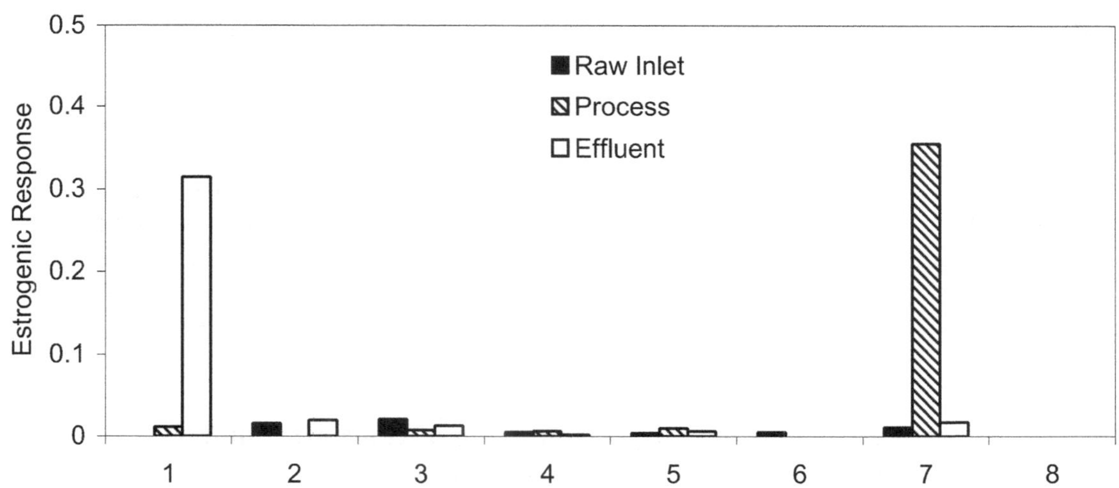


**Figure 7.2 YES response of unconcentrated water samples**

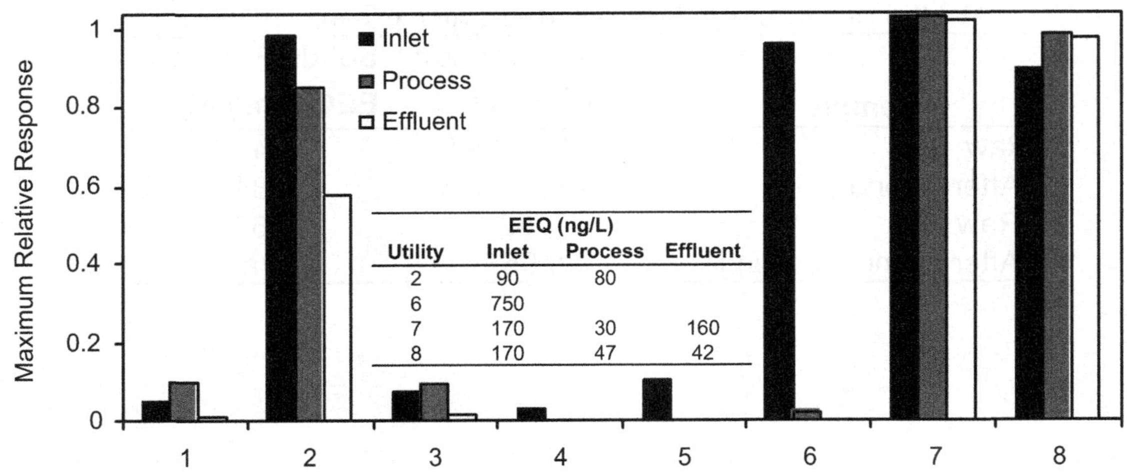


| | EEQ (ng/L) | | |
|---|---|---|---|
| Utility | Inlet | Process | Effluent |
| 2 | 90 | 80 | |
| 6 | 750 | | |
| 7 | 170 | 30 | 160 |
| 8 | 170 | 47 | 42 |

**Figure 7.3 Maximum relative YES response corresponding to 50x concentration of the water samples**

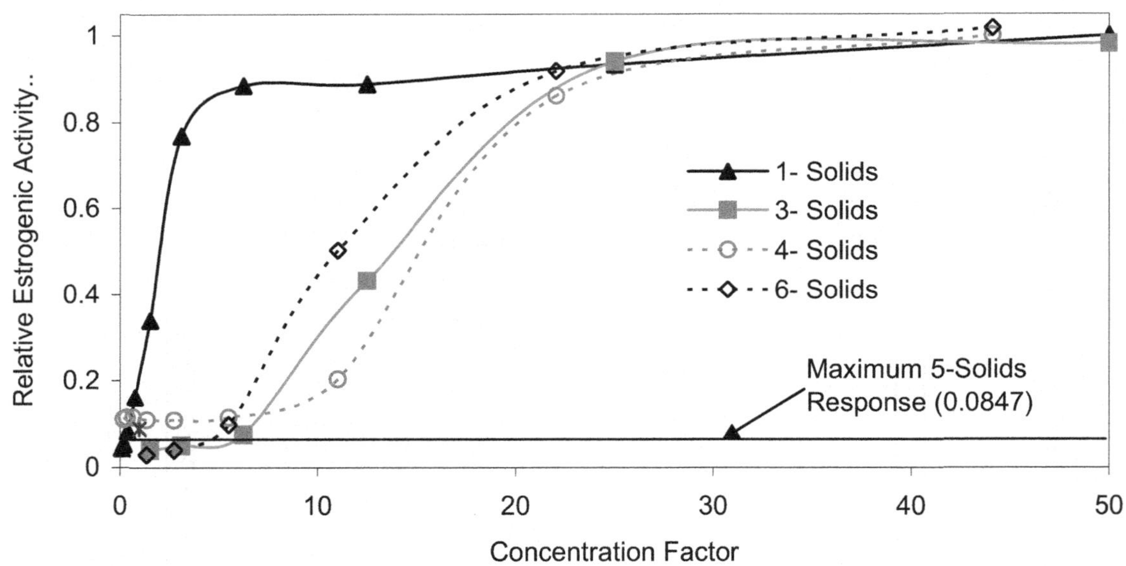


**Figure 7.4 Relative estrogenic activity as a function of concentration factor in the solids of five WTP (utilities 1, 3, 4, 5, 6) influent waters (concentration factor relative to whole water)**

# CHAPTER 8
# CONCLUSIONS AND RECOMMENDATIONS

## CONCLUSIONS

The kinetic results presented in this research show that UV technologies can be an effective treatment scheme for degradation of several EDCs in model and natural waters. The UV based destruction of EDCs can be effectively modeled, based on fundamental kinetic parameters in various water matrices, including natural, surface water of known pH, alkalinity, and TOC. By utilizing this modeling approach, along with known kinetic parameters such as the OH radical rate constant and quantum yield for a compound, it is possible to predict the effectiveness of UV treatment processes for destroying the contaminant of concern. Mineralization of the pollutant would require extended UV treatment times, and in many cases would not be practical. Thus, as with any oxidation process, breakdown products will be formed. The relative threats of these products, as compared to their parent compounds, must be assessed through toxicological testing.

Our studies demonstrate that *in vitro* and/or *in vivo* biological assessment of residual estrogenic activity in the treated solution is an ideal way to determine effective treatment endpoints and complements traditional chemical based assessments. Differences between *in vitro* and *in vivo* assay results highlight the differential sensitivities of these bioanalytical assays and emphasize the necessity to employ mechanistically diverse assays for assessment of EDC activity in treatment studies. Although LP or MP UV alone did not effectively degrade BPA, UV/AOP under adequate $H_2O_2$ and UV fluence was highly effective for removing aqueous estrogenic activity to below detectable levels. However, due to the high initial concentrations of BPA used for the study of estrogenic activity, the UV fluence levels utilized were 50 to 100 times higher than what is typically applied in disinfection practice and 5 to 10 times higher than standard UV/AOP applications. Thus, UV-based oxidation treatment for EDC contaminants at environmentally relevant concentrations was investigated.

The transformation of the steroidal EDCs 17-β-estradiol (E2) and 17-α-ethinyl estradiol (EE2) by direct UV photolysis and $UV/H_2O_2$ advanced oxidation was studied at environmentally relevant concentrations using the YES for analysis. First, experiments were performed to link the oxidation of E2 and EE2 with subsequent reduction in estrogenic activity. No statistically significant difference between removal rates for chemical assessments and bioanalytical (YES) assessments was observed, implying that the oxidation products of E2 and EE2 are not as estrogenically active (as measured by the YES) as the parent compounds. In laboratory and natural waters, 90 percent oxidation of these species occurred at a UV dose of less than 400 mJ $cm^{-2}$, implying these compounds are significantly degraded at much lower UV doses than previously thought, when considered at environmentally relevant levels. This is an important finding in contrast to the high UV doses indicated in work where contaminants were spiked in at micromolar levels.

Interestingly, combined (additive or synergistic) effects of *in vitro* and *in vivo* estrogenic activity were found in EDC mixtures compared to a single compound in solution. In addition, it appeared that the synergistic effect of *in vitro* estrogenic activity in EDC mixtures was more significant in natural water than that observed in laboratory water. $UV/H_2O_2$ treatment at $H_2O_2$ doses (10 ppm) and UV fluence (<1000 mJ $cm^{-2}$) relevant to water treatment oxidation processes was capable of decreasing *in vitro* and *in vivo* estrogenic activity for an EDC mixture at low concentrations. Thus it was demonstrated that the use of biological assessments provide a unique approach to evaluate the performance of $UV/H_2O_2$ AOPs for μg-ng $L^{-1}$ of EDC mixture contaminated water, an advantage over chemical analyses.

Water supplies entering treatment plants in a variety of watersheds were evaluated for estrogenic activity using the YES assay. For each sample tested, unconcentrated water samples exhibited little activity. However, when samples were concentrated, the results yielded some interesting observations, including an apparent correlation between the location of the water source within an urban to suburban area and increased estrogenic activity. Utilities with sources located in more rural or small town settings did not show positive responses to the YES. Estrogenic activity was found associated with all the solids in the inlet waters with the exception of one utility. This was expected due to the hydrophobic nature of many xenoestrogenic compounds, and a mass based estrogen equivalent concentration for the solids (mg $kg^{-1}$) was calculated.

## RECOMMENDATIONS

The recommendations from this research stem from the use of UV for treatment of estrogenic contaminants and the means by which those contaminants are monitored and measured before and after treatment.

UV light alone either from LP or MP sources will not be effective for degradation of the estrogenic chemicals 17-β-estradiol (E2), 17-α-ethinyl estradiol (EE2), nonlyphenol, or bisphenol A. This statement is true for UV as used with disinfection (UV doses from 20 to 200 mJ $cm^{-2}$) as well as UV at higher doses used in chemical treatment (UV doses from 200 to 1000 mJ $cm^{-2}$). Therefore, utilities should not expect any effects on estrogenic contaminants under UV disinfection conditions. For effective treatment of the EDCs studied, and likely others, UV can be used in combination with low levels of hydrogen peroxide to achieve advanced oxidation conditions where hydroxyl radical is an effective oxidant for destroying parent compounds and residual estrogenic activity. Hydrogen peroxide levels between 5 and 10 mg $L^{-1}$ appear to be effective when combined with UV doses of 200 to 500 mJ $cm^{-2}$ for effectively treating environmentally relevant levels of contaminants.

A battery of bioassays combined with chemical analysis is essential to examine the treatment endpoints. A strategy that encompasses the evaluation of aqueous biological activity via *in vitro* and *in vivo* bioassays for treatments of EDCs should be applied as a method to critically assess the efficacy of UV and other water and or wastewater treatment processes. At a minimum, the use of a screening assay, such as the Yeast Estrogen Screen (YES) or other mammalian cell based assay, should be used to complement any chemical analysis when studying treatment processes that result in transformation of an estrogenic parent contaminant to its known or unknown breakdown products. These assays can pick up residual estrogenic activity of any byproducts that are formed. These assays can also be used to monitor estrogenic activity when mixtures of pollutants are present and provide an end point marker for indicating effective treatment levels.

## PRACTICAL CONSIDERATIONS FROM THE RESEARCH

The following practical considerations for water utilities were outcomes of this research:

1. UV doses used for disinfection will not likely be effective at removing any estrogenic contaminant activity that may be present in a water supply.
2. UV irradiation at doses 10 to 100 times higher than used in disinfection will not likely be effective at removing any estrogenic contaminant activity that may be present in a water supply.
3. Use of UV irradiation in combination with hydrogen peroxide to generate OH radicals as part of an advanced oxidation process can be a very effective process to treat estrogenic contaminants.
4. There was no appreciable difference between LP and MP UV for OH radical based treatment of estrogenic contaminants.
5. Hydrogen peroxide doses do not need to be excessively high for treatment; levels between 2 and 10 mg $L^{-1}$ can lead to effective treatment endpoints.
6. UV doses used as part of an advanced oxidation process should be at least 200 and may need to be as high as 1000 mJ $cm^{-2}$ to achieve effective treatment of estrogenic contaminants.
7. Water quality has a large effect on the efficiency of the UV-AOP process. If the water contains high levels of OH radical scavengers, a high UV dose will be required. Scavengers include carbonate species, natural organic matter and other non-target pollutants.
8. The UV-AOP process may only partially remove estrogenicity from contaminated water, depending upon the matrix of pollutants and water quality present.
9. Toxicity assessment of specific byproducts was too challenging due to the low volume and concentration of byproducts generated, although this information would be helpful if these limitations could be overcome.
10. In our laboratory, the use of the mammalian E-screen assay proved to not be as sensitive or reliable as the YES screening assay.
11. The use of the YES was an effective and relatively high throughput means of determining estrogenic activity of treated or untreated water samples.
12. *In vitro* assays are reliable and sensitive. Particularly the YES assay is easily set up and could be used at a water utility with the purchase of minimum equipment.
13. The *in vivo* assay using the detection of female reproductive proteins in male Medaka fish was a helpful complement to the estrogenic screening assays, providing a uniquely different and complementary endpoint for determining estrogenic activity.
14. Some discrepancy was noted between *in vitro* and *in vivo* assays, with medaka VTG exhibiting a greater sensitivity than the YES assay. *In vivo* assay are thus an excellent adjunct to *in vitro* screening, but less practical for routine assessments.
15. The maintenance of a fish stock for running bioassays will likely not be practical for use at a water treatment plant; however, laboratories specializing in fish based toxicity testing could perform such analyses.

## FUTURE RESEARCH CONSIDERATIONS

Future research direction indicated by this study include:

1. Evaluation of new UV sources, such as pulsed UV systems for improvements on direct photolysis treatment of emerging contaminants.
2. Investigations with naturally occurring estrogenic contaminants and their detection with chemical analyses as well as estrogenic screening assays.
3. Determination of how effective other water treatment process are at removing environmental levels of estrogenic contaminants.
4. Large scale survey of the extent of estrogenic activity found at the inlets to water treatment plants utilizing estrogenic screening assays.
5. Further evaluating the impact of water quality on treatment objectives and estrogenic byproduct formation.
6. Identification of specific oxidation byproduct intermediates that retain estrogenic activity.

# REFERENCES


Andersen, H. R., A.M. Andersson, S.F. Arnold, H. Autrup, M. Barfoed, N.A. Beresford, P. Bjerregaard, L.B. Christiansen, B. Gissel, R. Hummel, E.B. Jorgensen, B. Korsgaard, R. Le Guevel, H. Leffers, J. McLachlan, A. Moller, J.B. Nielsen, N. Olea, A. Oles-Karasko, F. Pakdel, K.E. Pedersen, P. Perez, N.E. Skakkeboek, C. Sonnenschein and A.M. Soto. 1999. Comparison of short-term estrogenicity tests for identification of hormone-disrupting chemicals. *Environmental Health Perspectives*, 107 Suppl 1:89-108.

Ankley, G., E. Mihaich, R. Stahl, D. Tillitt, T. Colborn, S. McMaster, R. Miller, J. Bantle, P. Campbell, N. Denslow, R. Dickerson, L. Folmar, M. Fry, J. Giesy, L.E. Gray, P. Guiney, T. Hutchinson, S. Kennedy, V. Kramer, G LeBlanc, M. Mayes, A. Nimrod, R. Patino, R. Peterson, R. Purdy, R. Ringer, P. Thomas, L. Touart, G. Van der Kraak, and T. Zacharewski. 1998. Overview of a workshop on screening methods for detecting potential (anti-) estrogenic/ androgenic chemicals in wildlife. *Environmental Toxicology and Chemistry*, 17 (1): 68-87.

Arcand-Hoy, L.D., and W.H. Benson, 1998. Fish reproduction; An ecologically relevant indicator of endocrine disruption. *Environmental Toxicology and Chemistry*, 17 (1): 49-57.

Arukwe, A., F.R. Knudsen, and A. Goksoyr. 1997. Fish zona radiata (eggshell) protein: A sensitive biomarker for environmental estrogens. *Environmental Health Perspectives*, 105(4): 418-422.

Arukwe, A., S.W. Kullman, and D.E. Hinton. 2001. Differential biomarker gene and protein expressions in nonylphenol and estradiol-17 beta treated juvenile rainbow trout (Oncorhynchus mykiss). *Comparative Biochemistry and Physiology C- Toxicology and Pharmacology,* 129 (1): 1-10.

Atkinson, A. and D. Roy. 1995. *In vitro* conversion of environmental estrogenic chemical bisphenol A to DNA binding metabolite(s). *Biochem. Biophys. Res. Commun.*, 210: 424–433.

Baxendale J.H. and J.A. Wilson. 1957. The photolysis of hydrogen peroxide at high lamp intensities. *Trans Faraday Soc*, 54: 344-356.

Belfroid, AC; A. Van der Horst A.D. Vethaak, A.J. Schafer, G.B.J Rigs, J. Wegener, and W.P. Cofino. 1999. Analysis and occurrence of estrogenic hormones and their glucuronides in surface water and waste water in the Netherlands. *Science of the Total Environment*, 225 (1-2): 101-108.

Beltran, F.J., G. Ovejero, and B. Acedo. 1993. Direct, radical and competitive reactions in the ozonation of water micropollutants. *Water Research*, 27(6): 1013-1021.

Bolton J.R., and M.I. Stefan 2002. Fundamental photochemical approach to the concepts of fluence (UV dose) and electrical energy efficiency in photochemical degradation reactions. *Research on Chemical Intermediates*, 28 (7-9): 857-870.

Bolton, J.R., and K.G. Linden, 2003. Standardization of methods for fluence (UV Dose) determination in bench-scale UV experiments. *ASCE-J. Environ. Eng,* 129: 209-215.

Brix, R., S. Hvidt, and L. Carlsen, 2001. Solubility of nonylphenol and nonylphenol ethoxylates. On the possible role of micelles. *Chemosphere*, 44(4): 759-63.

Brotons, J.A., M.F. Oleaserrano, M. Villalobos, V. Pedraza, and N. Olea, 1995. Xenoestrogens released from lacquer coatings in food cans. *Environmental Health Perspectives*, 103(6): 608-612.

Buxton G.V., C.L. Greenstock, W.P. Helman, and A.B. Ross. 1988. Critical review of rate constants for reactions of hydrated electrons, hydrogen atoms and hydroxyl radicals (.OH/.O-) in aqueous solution. *Journal of Physical and Chemical Reference Data*, 17 (2): 513-886.

Cairns, W.L. 1996. Comparison of UV Disinfection Technologies Using Low Intensity Monochromatic and High Intensity Polychromatic UV Lamps. In *Proc. Disinfecting Wastewater for Discharge and Reuse - CP3601, March 17-20, 1996, Portland, OR.* Water Environment Federation.

Kwan, A. *et al.* 1996. Factors for Selection of a High Intensity UV Disinfection System for a Large Scale Application. In *Proc. Disinfecting Wastewater for Discharge and Reuse - CP3601, March 17-20, 1996, Portland, OR.* Water Environment Federation.

Carney, M. W., D.E. Hinton, and S.W. Kullman. 2005. Differential toxicity of naphthoic acid isomers in medaka (Oryzias latipes) embryos using a 96-well plate format. *In preparation*.

Cater, S. R., M.I. Stefan, J.R. Bolton, and A. Safarzadeh-Amiri. 2000. UV/$H_2O_2$ Treatment of Methyl tert-Butyl Ether in Contaminated Waters. *Environmental Science and Technology* 34 (4): 659-662.

Chen, P.-J., K.G. Linden, D.E. Hinton, S. Kashiwada, E.J. Rosenfeldt, and S.W. Kullman. 2006. Biological assessments of Bisphenol A degradation in water following direct photolysis and UV advanced oxidation. *Chemosphere Vol. 65:* 1094–1102.

Cheng, C.Y., and W.H. Ding. 2002. Determination of nonylphenol polyethoxylates in household detergents by high-performance liquid chromatography. *Journal of Chromatography A*, 968: 143-150.

Christiansen, L.B., M. Winther-Nielsen, and C. Helwig. 2002. Feminisation of fish. The effect of estrogenic compounds and their fate in sewage treatment plants and nature, Environmental Project No. 729; Danish Environmental Protection Agency. Available at: http://www.mst.dk/udgiv/publications/2002/87-7972-305-5/html/default_eng.htmhttp://www.mst.dk/udgiv/publications/2002/87-7972-305-5/html/default_eng.htm.

Clancy, J.L., Z. Bukhari, T.M. Hargy, J.R. Bolton, B.W. Dussert, and N.M. Marshall. 2000. Using UV to inactivate Cryptosporidium. *Journal of American Water Works Association*, 92 (9): 97-104.

Clark, L.B., R.T. Rosen, T.G. Hartman, B.J. Louis, I.H. Suffet, R.L. Lippincott, and J.D. Rosen. 1992. Determination of Alkylphenol ethoxylates and their acetic-acid derivatives in drinking-water by particle beam liquid-chromatography mass-spectrometry. *International Journal of Environmental Analytical Chemistry*, 47(3): 167-180.

CNN 2002. http://www.cnn.com/2002/TECH/science/04/16/frog.mutations.ap/index.html

Colborn, T., F.S. Vom Saal, and A.M. Soto. 1993. Developmental effects of endocrine disrupting chemicals in wildlife and humans. *Environmental Health Perspectives*, 101: 378–384.

Coleman H.M., E.J. Routledge, J.P. Sumpter, B.R. Eggins, and J.A. Byrne. 2004. Rapid loss of estrogenicity of steroid estrogens by UVA photolysis and photocatalysis over an immobilised titanium dioxide catalyst. *Water Research*, 38: 3233-324.

Cooke, J.B. and D.E. Hinton. 1999. Promotion by 17 beta-estradiol and beta-hexachlorocyclohexane of hepatocellular tumors in medaka, Oryzias latipes. *Aquatic Toxicology*, 45(2-3): 127-145.

Cooper, R. L., and R.J. Kavlock. 1997. Endocrine disruptors and reproductive development: a weight-of-evidence overview. *Journal of Endocrinology*. 152 (2): 159-166.

DeCaprio, A. P. 1999. The Toxicology of Hydroquinone — Relevance to Occupational and Environmental Exposure. *Critical Reviews in Toxicology*, 29(3): 283-330.

Fang, H., W. Tong, R. Perkins, A.M. Soto, N.V. Prechtl, and D.M. Sheehan. 2000. Quantitative comparisons of *in vitro* assays for estrogenic activities. *Environmental Health Perspectives*, 108 (8): 723-729.

Folmar, L.C., N.D. Denslow, V. Rao, M. Chow, D.A. Crain, J. Enblom, J. Marcino, and L.J. Guillette. 1996. Vitellogenin induction and reduced serum testosterone concentrations in feral male carp (Cyprinus carpio) captured near a major metropolitan sewage treatment plant. *Environmental Health Perspectives*, 104(10): 1096-1101.

Fry D.M., and C.K. Toone. 1981. "DDT – Induced Feminization of Gull Embryos" *Science*, 213 (4510): 922-924.

Fujii K.; S. Kikuchi, M. Satomi, N. Ushio-Sata, and N. Morita. 2002. Degradation of 17ß-Estradiol by a Gram-Negative Bacterium Isolated from Activated Sludge in a Sewage Treatment Plant in Tokyo, Japan. *Applied and Environmental Microbiology*, 68 (4): 2057-2060.

Fukahori, S., H. Ichiura, T. Kitaoka, and H. Tanaka. 2003. Capturing of bisphenol A photodecomposition intermediates by composite TiO2–zeolite sheets. *Applied Catalysis B-Environmental*, 46: 453-462.

Garcia-Reyero N., E. Grau, M. Castillo, M.J.L. De Alda, D. Barcelo, and B. Pina. 2001. *Environmental Toxicology and Chemistry*, 20 (6): 1152-1158.

Glaze W.H., Y. Lay, and J.W. Kang. 1995. Advanced Oxidation Processes – A kinetic model for the oxidation of 1,2-Dibromo-3-chloropropane in water by the combination of hydrogen peroxide and UV radiation. *Industrial and Engineering Chemical Research*, 34 (7): 2314-2323.

Gomes, R. L., M.D. Scrimshaw, and J.N. Lester. 2003. Determination of endocrine disrupters in sewage treatment and receiving waters. *Trends in Analytical Chemistry*, 22 (10): 697-707.

González-Doncel, M., E. de la Peña, C. Barrueco, and D.E. Hinton. 2003. Stage sensitivity of medaka (Oryzias latipes) eggs and embryos to permethrin. *Aquatic Toxicology*, 62 (3): 255-268.

Gonzalez-Doncel M, Fernandez-Torija C, Hinton DE, Tarazona JV. 2005. Stage-specific toxicity of cypermethrin to medaka (Oryzias latipes) eggs and embryos using a refined methodology for an in vitro fertilization bioassay. Arch Environ Contam Toxicol 48:87-98

Goodbred SL, R.J. Gilliom, T.S. Gross, N.P. Denslow, W.L. Bryant, and T.R. Schoeb. 1997. Report 96-627. U.S. Geological Survey, Denver, CO.

Gray, L.E., J. Ostby, V. Wilson, C. Lambright, K. Bobseine, P. Hartig, A. Hotchkiss, C. Wolf, J. Furr, M. Price, L. Parks, R.L. Cooper, T.E. Stoker, S.C. Laws, S.J. Degitz, K.J. Jensen, M.D. Kahl, J.J. Korte, E.A. Makynen, J.E. Tietge, and G.T. Ankley. 2002. Xenoendocrine disrupters-tiered screening and testing. *Toxicology*, 27: 371–382.

Guillette L.J., T.S. Gross, G.R. Masson, J.M. Matter, H.F. Percival, and A.R. Woodward. 1994. Developmental Abnormalities of the Gonad and Abnormal Sex-Hormone Concentrations in Juvenile Alligators from Contaminated and Control Lakes in Florida. *Environmental Health Perspectives*, 102(8): 680-699.

Hamazaki, T.S., I. Iuchi and K. Yamagami. 1987. Purification and identification of vitellogenin and its immunohistochemical detection in growing oocytes of the teleost, Oryzias latipes. *Journal of Experimental Zoology*, 242: 333–341.

Hames, B. D. 1998. *Gel Electrophoresis of Protein: A Practical Approach.* New York: Oxford University Press. pp46; pp53-78.

Hayes T.B., A. Collins, M. Lee, M. Mendoza, N. Noriega, A.A. Stuart, and A. Vonk. 2002. Hermaphroditic, Demasculinized Frogs after Exposure to the Herbicide Atrazine at Low Ecologically Relevant Doses. *Proceedings of the National Academy of Sciences of the United States of America*, 99 (8): 5476-5480.

Hu J.Y., T. Aizawa, and S. Ookubo. 2002a. Products of Aqueous Chlorination of Bisphenol A and their Estrogenic Activity. *Environmental Science and Technology*, 36 (9): 1980-1987.

Hu J.Y., G.H. Xie, and T. Aizawa. 2002b. Products of Aqueous Chlorination of 4-nonylphenol and their Estrogenic Activity. *Environmental Toxicology and Chemistry*, 21 (10): 2034-2039.

Huber M.M., S. Canonica, G.Y. Park, and U. von Gunten. 2003. Oxidation of Pharmaceuticals during ozonation and advanced oxidation processes. *Environmental Science and Technology*, 37 (5): 1016-1024.

Huber M.M., S. Korhonen, T.A. Ternes, and U. von Gunten. 2005. Oxidation of pharmaceuticals during water treatment with chlorine dioxide. *Water Research*, 39: 3607-3617.

Huber, M. M.; T.A. Ternes, and U. von Gunten. 2004. Removal of estrogenic activity and formation of oxidation products during ozonation of 17α-ethinylestradiol. *Environmental Science and Technology*, *38* (19): 5177-5186.

Huggett, D.B., C. M. Foran, B. W. Brooks, J. Weston, B. Peterson, K. E. Marsh, T. W. L. Point, and D. Shlenk. 2003. Comparison of *in vitro* and *in vivo* Bioassays for Estrogenicity in Effluent from North American Municipal Wastewater Facilities. *Toxicological Sciences*, 72: 77-83.

Ike, M., M.Y. Chen, C.S. and Jin, M.M. Fujita. 2002. Acute toxicity, mutagenicity, and estrogenicity of biodegradation products of bisphenol-A. *Environmental Toxicology*, 17: 457-461.

Irmak S., O. Erbatur, and A. Akgerman. 2005. Degradation of 17β-estradiol and bisphenol A in aqueous medium by using ozone and ozone/UV techniques. *Journal of Hazardous Materials B*, 126: 54-62.

Ishikawa, Y. 2000. Medakafish as a model system for vertebrate developmental genetics. *Bioassays*, 22: 487-495.

Iwamatsu, T. 1994. Stages of normal development in the medaka. *Oryzias latipes. Mech Dev.*, 121 (7-8): 605-618.

Jobling S., D. Sheahan, J.A. Osborne, P. Matthiessen, and J.P. Sumpter, 1996. Inhibition of Testicular Growth in Rainbow Trout (Oncorhynchus mikiss) Exposed to Estrogenic Alkylphenolic Chemicals *Environmental Toxicology and Chemistry*, 15(2): 194.

Jobling, S., M. Nolan, C.R. Tyler, G. Brighty, and J.P. Sumpter. 1998. Widespread sexual disruption in wild fish. *Environmental Science and Technology*, 32(17):2498-2506.

Jorgensen, M., B. Vendelbo, N.E. Skakkebaek, and H. Leffers. 2000. Assaying estrogenicity by quantitating the expression levels of endogenous estrogen-regulated genes. *Environmental Health Perspectives*, 108 (5): 403-412.

Jurgens MD, K.I.E. Holthaus, A.C. Johnson, J.J.L. Smith, M. Hetheridge, and R.J. Williams. 2002. The potential for Estradiol and Ethinylestradiol degradation in English rivers. *Environmental Toxicology and Chemistry*, 21(3): 480-488.

Kabuto, H., M. Amakawa, and T. Shishibori. 2004. Exposure to bisphenol A during embryonic / fetal life and infancy increases oxidative injury and causes underdevelopment of the brain and testis in mice. *Life Science*, 74: 2931-2940.

Kang J.H., and F. Kono. 2002. Effects of Bacterial Counts and Temperature on the Biodegradation of Bisphenol A in River Water. *Chemosphere*, 49 (5): 493.

Kashiwada, S., H. Ishikawa, N. Miyamoto, Y. Ohnishi, and Y. Magara. 2002. Fish test for endocrine-disruption and estimation of water quality of Japanese rivers. *Water Research*, 36: 2161-2166.

Kelce, W.R.; C.R. Stone, S.C. Laws, L.E. Gray, and J.A. Kemppainen, and E.M. Wilson. 1995. Persistent DDT metabolite p,p'-DDE is a potent androgen receptor antagonist. *Nature*, 375 (6532): 581-585.

Kimura K, S. Toshima, G. Amy, and Y. Watanabe 2004. Rejection of neutral endocrine disrupting compounds (EDCs) and pharmaceutical active compounds (PhACs) by RO membranes. *Journal of Membrane Science*, 245: 71-78.

Klassen, N.V., D. Marchington, and H.C.E. McGowan. 1994. $H_2O_2$ determination by the $I_3^-$ method and by $KMnO_4$ titration. *Analytical Chemistry*, 66: 2921-2925.

Koger, C. S., and D.E. Hinton. 2000. Determining the sensitive developmental stages of intersex induction in medaka (Oryzias latipes) exposed to 17β-estradiol or testosterone. *Marine Environmental Research*, 50: 201–206.

Kolpin D.C., E.T. Furlong, M.T. Meyer, M. Thurman, S.D. Zaugg, L.B. Carber, and H.T. Buston. 2002. Pharmaceuticals, hormones, and other organic wastewater contaminants in US streams, 1999-2000: A national reconnaissance. *Environmental Science and Technology*, 36(6): 1202-1211.

Korner, W., W. Boltz, W. Submuth, G. Hiller, V. Hanf, and H. Hagenmaier. 2000. Input/output balance of estrogenic active compounds in a major municipal sewage plant in Germany. *Organohalogen Comp*. 37: 269-272.

Kwan, A., J. Archer, F. Soroushian, A. Mohammed, and G. Tchobanoglous. 1996. Factors for Selection of a High Intensity UV Disinfection System for a Large Scale Application. In Proc. Disinfecting Wastewater for Discharge and Reuse - CP3601, March 17-20, 1996, Portland, OR. Water Environment Federation.

Lean, G., and R. Sadler. 2002. Male fertility fears over pollution in water supply. *The Independent*, March 17, 2002. London, England.

Larson R, and R. Zepp. 1988. Reactivity of the Carbonate Radical with Aniline Derivatives. *Environmental Toxicology and Chemistry*, 7 (4): 265-274.

Lenz K., V. Beck, and M. Fuerhacker. 2004. Behavior of bisphenol A (BPA), 4-nonylphenol (4-NP), and 4-nonylphenol ethoxylates (4-NP1EO, 4-NP2EO) in oxidative water treatment processes. *Water Science and Technology*, 50 (5): 141-147.

Liu, W., S.A. Andrews, M.I. Stefan, and J.R. Bolton. 2003. Optimal methods for quenching H2O2 residuals prior to UFC testing. *Water Research*, 37 (15): 3697-3703.

Merril, C. R., and K.M. Washart. 1998. Protein detection methods. In *Hames, B.D. (Eds). Gel Electrophoresis.* New York: Oxford University Press. 53-78.

Morse, D.C., E.K. Wehler, W. Wesseling, J.H. Koeman, and A. Brouwer. 1996. Alterations in rat brain thyroid hormone status following pre- and postnatal exposure to polychlorinated biphenyls (Aroclor 1254). *Toxicology and Applied Pharmacology*, 136 (2): 269-279.

Nakashima T, Y. Ohko, D.A. Tryk, and A. Fujishima. 2002. Decomposition of endocrine-disrupting chemicals in water by use of TiO2 photocatalysts immobilized on polytetrafluoroethylene mesh sheets. *Journal of Photochemistry and Photobiology A – Chemistry*, 151 (1-3): 207-212.

Nasu M., M. Goto, H. Kato, Y. Oshima, and H. Tanaka. 2001. Study on endocrine disrupting chemicals in wastewater treatment plants. *Water Science and Technology*, 43(2): 101-108.

Nghiem L.D., A. Manis, K. Soldenhoff, and A.I. Schafer. 2004. Estrogenic hormone removal from wastewater using NF/RO membranes. *Journal of Membrane Science*, 242: 37-45.

Nilsen B.M., K. Berg, J.K. Eidem, S.I. Kristiansen, F. Brion, J.M. Porcher, and A. Goksoyr. 2004. Development of quantitative vitellogenin-ELISAs for fish test species used in endocrine disruptor screening. *Anal Bioanalytical Chem*istry, 378 (3): 621-633.

Nishi K., M. Chikae, Y. Hatano, H. Mizukami, M. Yamashita, R. Sakakibara, and E. Tamiya. 2002. Development and application of a monoclonal antibody-based sandwich ELISA for quantification of Japanese medaka (Oryzias latipes) vitellogenin. *Comp Biochem Physiol C Toxicol Pharmacol*, 132 (2): 161-169.

O'Connor, J.C., J.C. Cook, M.S. Marty, L.G. Davis, A.M. Kaplan, and E.W. Carney. 2002. Evaluation of tier I screening approaches for detecting endocrine-active compounds (EACs). *Crit. Rev. Toxicol.*, 32: 521–549.

Ohko Y., I. Ando, C. Niwa, T. Tatsuma, T. Yamamura, T. Nakashima, Y. Kubota, and A. Fujishima. 2001. Degradation of Bisphenol A in Water by $TiO_2$ Photocatalyst. *Environmental Science and Technology*, 35 (11): 2365-2368.

Ohko Y, K.I. Iuchi, C. Niwa, T. Tatsuma, T. Nakashima, T. Iguchi, Y. Kubota, and A Fujishima. 2002. 17-β-Estradiol degradation by TiO2 photocatalysis as means of reducing estrogenic activity. *Environmental Science and Technology*, 369 (19): 4175-4181.

Planas C., J.M. Guadayol, M. Droguet, A. Escalas, J. Rivera, and J. Caixach. 2002. Degradation of polyethoxylated nonylphenols in a sewage treatment plant. Quantitative analysis by isotopic dilution-HRGC/MS. *Water Research,* 36 (4): 982-988.

Papoulias, D.M., D.R. Buckler, and D.E. Tillit. 1996. Optimization of the Ames/Salmonella mutagenicity assay for use with extracts of aquatic sediments. *Journal of Great Lakes Research*, 22 (3): 584-590.

Papoulias, D.M., D.B. Noltie, and D.E. Tillit. 2000. An *in vivo* model fish system to test chemical effects on sexual differentiation and development: exposure to ethinyl estradiol. *Aquatic Toxicology*, 48 (1): 37-50.

Purdom, C.E., P.A. Hardiman, V.J. Bye, N.C. Eno, C.R. Tyler, and J.P. Sumpter. 1994. Estrogenic effects of effluents from sewage treatment works. Chem. Ecol. 8: 275-285.

Rosenfeldt E.J., and K.G. Linden. 2004. Degradation of endocrine disrupting chemicals bisphenol A, ethinyl estradiol, and estradiol during UV photolysis and advanced oxidation processes. *Environmental Science and Technology*, 38: 5476-5483.

Rosenfeldt EJ, and K.G. Linden. 2006. The $R_{OH,UV}$ concept to characterize and model UV/H2O2 process in natural water. *submitted to Environmental Science and Technology.*

Rosenfeldt, E.J., B. Melcher, and K.G. Linden. 2005. Treatment of Taste and Odor Causing Compounds in Water by UV and UV/H2O2 Processes. *Journal of Water Supply: Research & Technology –AQUA*, 54(7): 423-434.

Routledge, E., J.P. Sumpter. 1996. Estrogenic activity of surfactants and some of their degradation products assessed using a recombinant yeast screen. *Environmental Toxicology and Chemistry*, 15 (3): 241–248.

Sapozhnikova, Y., A. McElroy, S. Snyder, and D. Schlenk. 2004. Estrogenic Activity measurement in Wastewater using *in vitro* and *in vivo* Methods. *Personal communications*.

Schultz, T.W., D.H. Kraut, G.S. Sayler, and A.C. Layton. 1998. Estrogenicity of selected biphenyls evaluated using a recombinant yeast assay. *Environmental Toxicology and Chemistry*, 17 (9): 727-729.

Schwarzenbach, R.P., P.M. Gschwend, and D.M. Imboden. 1993. *Environmental Organic Chemistry*; New York: Wiley-Interscience, John Wiley and Sons. Chapter 13.

Segner H, Caroll K, Fenske M, Janssen CR, Maack G, Pascoe D, et al. 2003. Identification of endocrine-disrupting effects in aquatic vertebrates and invertebrates: report from the European IDEA project. Ecotoxicol Environ Safe 54:302-314.

Sharpless C.M. and K.G. Linden. 2003. Experimental and Model Comparisons of Low- and Medium-Pressure Hg Lamps for the Direct and $H_2O_2$ Assisted UV Photodegradation of *N*-Nitrosodimethylamine in Simulated Drinking Water. *Environmental Science and Technology*, 37 (9): 1933-1940.

Sharpless C.M. and K.G. Linden. 2001. UV Photolysis of Nitrate: Effects of Natural Organic Matter and Dissolved Inorganic Carbon and Implications for UV Water Disinfection. *Environmental Science and Technology*, 35: 2949-2955.

Shin G.A., K.G. Linden, M. Arrowood, and M.D. Sobsey. 2001. Low pressure UV inactivation and subsequent DNA repair potential of *Cryptosporidium parvum* oocysts. *Applied Environmental Microbiology*, 67 (7): 3029-3032.

Silva, E., N. Rajapakse, and A. Kortenkamp. 2002. Something from "nothing" - eight weak estrogenic chemicals combined at concentrations below NOECs produce significant mixture effects. *Environmental Science and Technology* 36(8): 1751-1756.

Skehan P, Storeng R, Scudiero D, Monks A, McMahon J, Vistica D, Warren JT, Bokesch H, Kenney S, Boyd MR. (1990). New colorimetric cytotoxicity assay for anticancer-drug screening. J. Natl. Cancer Inst. 82, 1107-1112.

Snyder S.A., P. Westerhoff, Y. Yoon, and D.L. Sedlack. 2003. Pharmaceuticals, Personal Care Products, and Endocrine Disruptors in Water: Implications for the Water Industry. *Environmental Engineering Science* 20 (5): 449-469.

Solomon, G. and T. Schettler. 2000. Environment and Health: 6. Endocrine disruption and potential human health implications. *Canadian Medical Association Journal*, 163: 1471-1476.

Sonnenschein, C. and A.M. Soto. 1998. An updated review of environmental estrogen and androgen mimics and antagonists. *Journal of Steroid Biochemistry,* 65: 143-150.

Soto A. M., C. Sonnenschein, K. L. Chung, M. F. Fernandez, N. Olea, and F. O. Serrano. 1995. The E-SCREEN assay as a tool to identify estrogens: an update on estrogenic environmental pollutants. *Environmental Health Perspectives,* 103 (Suppl 7): 113-22.

Staples, C.A., P.B. Dorn, G.M. Klecka, S.T. O'Block, and L.R. Harris. 1998. A review of the environmental fate, effects, and exposures of bisphenol A. *Chemosphere*, 36: 2149-2173.

Stefan, M.I., and J.R. Bolton. 1998. Mechanism of the degradation of 1,4-dioxane in dilute aqueous solution using the UV hydrogen peroxide process. *Environmental Science and Technology,* 32 (11): 1588-1595.

Stefan, M.I., J. Mack, and J.R. Bolton. 2000. Degradation pathways during the treatment of methyl tert-butyl ether by the UV/$H_2O_2$ process. *Environmental Science and Technology*, 34(4): 650-658.

Stefan, M. I., C.M. Sharpless, K.G. Linden, and J.R. Bolton. 2001. Degradation of Atrazine in a Bench-scale Collimated Beam Apparatus Using Low- and Medium-Pressure UV Lamps. In *Proceedings, American Water Works Association Water Quality and Technology Conference, Nashville, TN.* Denver, Colo: AWWA.

Sumpter, J. P., and S. Jobling. 1995. VTG Vitellogenesis as a biomarker for estrogenic contamination of the aquatic environment. *Environmental Health Perspectives* 103 (Suppl 7): 173-178.

Tabata A., N. Miyamoto, Y. Ohnishi, M. Itoh, T. Yamada, T. Kamei, and Y. Magara. 2003. The effect of chlorination of estrogenic chemicals on the level of serum vitellogenin of Japanese medaka (*Oryzias latipes*). *Water Science and Technology*, 47: 51-57.

Tanaka H. Y., A. Yakou, T. Takahashi, T. Higashitani, and K. Komori. 2001. Comparison between estrogenicities estimated from DNA recombinant yeast assay and from chemical analyses of endocrine disruptors during sewage treatment. *Water Science and Technology*, 43 (2): 125-132.

Teh, S.J., and D.E. Hinton. 1998. Gender-specific growth and hepatic neoplasia in medaka (Oryzias latipes). *Aquatic Toxicology* 41 (1-2): 141-159.

Thorpe, K. L., and R.I. Cummings, T.H. Hutchinson, M. Scholze, G. Brighty, J.P. Sumpter, and C.R. Tyler. 2003. Relative Potencies and Combination Effects of Steroidal Estrogens in Fish. *Environmental Science and Technology*, *37* (6): 1142-1149.

Toomey, B.H., G.H. Monteverdi, and R.T. Di Giulio. 1999. Oxtylphenol induces vitellogenin production and cell death in fish hepatocytes. *Environmental Toxicology and Chemistry*, 18 (4): 734-739.

USEPA. 1991. *Technical Support Document for Water Quality-based Toxics Control.* Office of Water. Washington, DC. EPA/505/2-90-001.

USEPA. 1993a. *Methods for Aquatic Toxicity Identification Evaluations: Phase II Toxicity Identification Procedures for Samples Exhibiting Acute and Chronic Toxicity.* Office of Research and Development. Washington, DC. EPA/600/R-92-080.

USEPA. 1993b. *Methods for Aquatic Toxicity Identification Evaluations: Phase III Toxicity Identification Procedures for Acutely and Chronically Toxic Samples.* U.S. Environmental Protection Agency, Duluth, MN. EPA/600/R-92-081.

USEPA. 2000. *Understanding and accounting for method variability in whole effluent toxicity applications under the National Pollutant Discharge Elimination System Program.* Denton DL, Fox J, Fulk FA, Greenwald K, Narvaez M, Norberg-King TJ, Phillips L. eds. Office of Water. Washington, DC. EPA/833/R-00-003.

USEPA. 2001. *Removal of Endorine Disruptor Chemicals Using Drinking Water Treatment Processes*. Washington, DC. EPA/625/R-00/015.

Van den Belt , K., P. Berckmans, C. Vangenechten, R. Verheyen, and H. Witters. 2004. Comparative study on the *in vitro/in vivo* estrogenic potencies of 17β-estradiol, estrone, 17α-ethynylestradiol and nonylphenol. *Aquatic Toxicology*, 66 (2): 183-195.

Villalobos S.A., J.T. Hamm, S.J. Teh, D.E. Hinton. 2000 Thiobencarb-induced embryotoxicity in medaka (Oryzias latipes): stage-specific toxicity and the protective role of chorion. Aquat Toxicol. 48(2-3):309-326.

Weimeyer S.N., T.G. Lamont, C.M. Bunk, C.R. Sindelar, F.J. Gramlich, J.D. Frasher, and M.A. Byrd. 1984. Organochlorine, Pesticide, Polychlorobiphenyl, and Mercury Residues in Bald Eagle Eggs – 1969 – 79 – And Their Relationships to Shell Thinning and Reproduction. *Archives of Environmental Contamination and Toxicology,* 13 (5): 529.

Welch, R. M., W. Levin, and A. H. Conney. 1969. Estrogenic action of DDT and its analogs. *Toxicology and Applied Pharmacology,* 14 (2):358-367.

Wells C.H.J. 1972. In *Introduction to Molecular Photochemistry.* London: Chapman and Hall, Ltd.

Westerhoff P., Y. Yoon, S. Shyder, and E. Wert. 2005. Fate of Endocrine-Disruptor, Pharmaceutical, and Personal Care Product Chemicals during Simulated Drinking water treatment processes. *Environmental Science and Technology*, 39: 6649-6663.

Yamamoto, T. 1975. *Stages in the development. Medaka (Killifish). Biology and Strains*. Tokyo: Keigaku Publ Co. pp: 59-72.

Yamamoto, T., A. Yasuhara, H. Shiraishi, and O. Nakasugi. 2001. Bisphenol A in hazardous waste landfill leachates. *Chemosphere,* 42: 415-418.

Ying G.G., R.S. Kookana, and Z.L. Chen. 2002. On-line solid-phase extraction and fluorescence detection of selected endocrine disrupting chemicals in water by high-performance liquid chromatography. *Journal of Environmental Science and Health Part B – Pesticides Food Contaminants and Agricultural Wastes*, 37 (3): 225-234.

Yoon Y., P. Westerhoff, and S.A. Snyder. 2005. Adsorption of 3H-labeled 17-β-estradiol on powdered activated carbon. *Water, Air, and Soil Pollution*, 166: 343-351.

Yoon Y., P. Westerhoff, J. Yoon, and S.A. Snyder. 2004. Removal of 17β Estradiol and Fluoranthene by Nanofiltration and Ultrafiltration. *Journal of Environmental Engineering,* 130 (12): 1460-1467.

Yanping Z. and J.L. Zhou. 2005. Removal of estrone and 17β-estradiol from water by adsorption. *Water Research,* 39: 3991-4003.

Zepp R.G., J. Hoigne, and H. Bader. 1987. Nitrate-induced photooxidation of trace organic chemicals in water. *Environmental Science and Technology, 21: 443-450.*

Zwiener, C., and F.H. Frimmel. 2000. Oxidative treatment of pharmaceuticals in water *Water Research,* 34 (6): 1881-1885.

# ABBREVIATIONS AND SYMBOLS

| ABBREVIATION | DEFINITION |
|---|---|
| ACN | acetonitrile |
| AOP | advanced oxidation process |
| AOT | advanced oxidation technology |
| AwwaRF | Awwa Research Foundation |
| BPA | bisphenol A |
| CI | confidence interval |
| cm | centimeter(s) |
| DOC | dissolved organic carbon |
| E | Irradiance ($mW/cm^2$) |
| E2 | estradiol |
| $EC_{50}$ | effective concentrations to half maximum response |
| EE2 | ethinyl estradiol |
| EDC | Endocrine Disrupting Contaminant |
| EEF | Estradiol equivalency factor |
| EEQ | $E_2$ equivalent concentrations |
| ELISA | Enzyme-Linked Immunosorbent Assay |
| ER | estrogen receptor |
| ERE | estrogen response element |
| ft | foot (feet) |
| g | gram |
| hER | human estrogen receptor |
| $H_{200\text{-}300}$ | UV dose from 200 to 300 nm ($mJ/cm^2$) |
| $H_2O_2$ | hydrogen peroxide |
| hr(s) | hour(s) |
| HPLC | High Performance Liquid Chromatography |
| in. | inch(es) |
| J | joule(s) |
| kDa | kilodaltons |
| kW | kilowatt |
| L | liter(s) |
| LC | liquid chromatography |

| | |
|---|---|
| $LC_{50}$ | lethal concentration at half maximum mortality |
| LP | low-pressure |
| LPHO | low-pressure high output |
| LT2ESWTR | long-term 2 ESWTR |
| | |
| m | meter(s) |
| µm | micrometer(s) |
| µs | microsecond |
| µW | microwatt(s) |
| mM | millimolar |
| mgd, MGD | million gallon(s) per day |
| mg/L | milligram(s) per liter |
| mJ | millijoule(s) |
| min | minutes |
| mL | milliliter(s) |
| mm | millimiter(s) |
| MP | medium-pressure |
| mW | milliwatt(s) |
| | |
| n | number of samples |
| NDMA | n-nitrosodimethylamine |
| NOEC | non-observed-effected concentration |
| NOM | natural organic matter |
| NP | nonylphenol |
| nm | nanometer(s) |
| ntu, NTU | nephelometric turbidity unit(s) |
| | |
| OD | optical density |
| OH | hydroxyl radical |
| ONPG | ortho-nitrophenyl-beta-D-galactopyranoside |
| | |
| pCBA | para-Chloro-Benzoic Acid |
| PE | proliferation effect |
| pH | negative logarithm of the effective hydrogen-ion concentration |
| PPCP | pharmaceutical and personal care products |
| ppm | parts per million |
| P-UV | Pulsed UV |
| | |
| R | correlation coefficient |
| | |
| s | second(s) |
| sq. cm | square centimeter(s) |
| SPE | Solid Phase Extraction |
| | |
| t | time |
| TOC | total organic carbon |

| | |
|---|---|
| TSK | Trimmed Spearman-Karber |
| U.S. | United States |
| U.S. EPA | United States Environmental Protection Agency |
| UV | ultraviolet |
| $UV_{254}$ | Absorbance of 254-nanometer ultraviolet light |
| VTG | Vitellogenin |
| W | watt(s) |
| WTP | water treatment plant |
| YES | yeast estrogen screen |
| YNB | yeast nitrogen base |

| **SYMBOL** | **DEFINITION** |
|---|---|
| $^{o}C$ | degree(s) Celsius |
| $^{o}F$ | degree(s) Fahrenheit |
| % | percent |